FISHERIES MATHEMATICS

FISHERIES MATHEMATICS

The Proceedings of a Conference organized by the Institute of Mathematics and its Applications, held 24–26 November, 1975, at the Marine Laboratory, Aberdeen

Edited by

Dr. J. H. STEELE

Marine Laboratory, Aberdeen
Scotland

UNIV. COLL. N.W. BANGOR LIBRARY

1977

ACADEMIC PRESS

LONDON · NEW YORK · SAN FRANCISCO

A Subsidiary of Harcourt Brace Jovanovich, Publishers

ACADEMIC PRESS INC. (LONDON) LTD.
24/28 Oval Road
London NW1

United States Edition published by
ACADEMIC PRESS INC.
111 Fifth Avenue
New York, New York 10003

781099
SH331.5.M48

Copyright © 1977 by
Institute of Mathematics and its Applications

All Rights Reserved
No part of this book may be reproduced in any form by photostat, microfilm or any other means, without written permission from the publishers

Library of Congress Catalog Card Number: 77-71838
ISBN 0-12-665250-3

Printed in Great Britain by Whitstable Litho Ltd., Whitstable, Kent

The Institute of Mathematics and its Applications
records its thanks to the editor, Dr. J. H. Steele, and
to the authors for their contributions to this volume

CONTRIBUTORS

K. P. Anderson, *Danish Institute for Fishery and Marine Research, Charlottenlund, Denmark.*

D. H. Cushing, *Fisheries Laboratory, Lowestoft, England.*

R. S. T. Ferro, *Marine Laboratory, Aberdeen, Scotland.*

J. J. Foster, *Marine Laboratory, Aberdeen, Scotland.*

F. R. Harden Jones, *Fisheries Laboratory, Lowestoft, England.*

E. W. Henderson, *Marine Laboratory, Aberdeen, Scotland.*

J. W. Horwood, *Fisheries Laboratory, Lowestoft, England.*

C. Johnston, *Marine Laboratory, Aberdeen, Scotland.*

R. Jones, *Marine Laboratory, Aberdeen, Scotland.*

Sir James Lighthill, *University of Cambridge, England.*

M. D. Nicholson, *Marine Laboratory, Aberdeen, Scotland.*

J. A. Pope, *Marine Laboratory, Aberdeen, Scotland.*

J. G. Pope, *Fisheries Laboratory, Lowestoft, England.*

A. J. Reid, *Marine Laboratory, Aberdeen, Scotland.*

J. B. Richardson, *White Fish Authority, Hull, England.*

J. H. Steele, *Marine Laboratory, Aberdeen, Scotland.*

E. Ursin, *Danish Institute for Fishery and Marine Research, Charlottenlund, Denmark.*

C. S. Wardle, *Marine Laboratory, Aberdeen, Scotland.*

Foreword

This book on "Fisheries Mathematics" is likely to be of wide interest to two groups of readers. Life scientists, and also engineers and mathematicians, who are responsible for advising government or industry on fisheries matters, will find here a broad compendium of modern approaches to the quantitative handling of the available data in their field. At the same time, applied mathematicians in general (whether their central expertise is in statistics or in mechanics) will find in this book an introduction to one of the most fascinating, modern areas of application of mathematics.

Every chapter of the book has much to interest both groups of readers. The book begins, as it should, with a discussion of primary production in the ocean, how the microscopic plants called phytoplankton "bloom" in the spring and how their rapid growth is later brought under control by two effects: limited availability of nutrients and grazing by herbivorous zooplankton (which, themselves, are the food of fishes such as herring). Standard mathematical representations of this process are briefly described, with full references.

Those representations ignore, however, the observed "patchiness" of the phytoplankton distribution. This "patchiness" is strongly aggravated by spatial variations in the moment of onset of the phytoplankton bloom and, conversely, may be reduced through mixing due to water turbulence. Readers already concerned with the population dynamics underlying fisheries science will be impressed by the clear evidence in Chapter 1 that patchiness is extremely marked and substantially affects those population dynamics. Applied mathematicians with an interest in the statistics (or in the mechanics) of turbulence will be interested to see the method of the *spatial spectrum* (much used in turbulence research) applied here to analyse plankton patchiness. At lower length-scales, control by turbulent mixing is evident from the fact that the plankton spectrum and turbulence spectrum are very similar. At higher length-scales, the growth properties and other aspects of the population dynamics of the plankton are in control, and can strongly enhance any initial patchiness. Chapter 1 makes a fascinating first attempt at a fully interactive theory of these phenomena.

From primary production, the book moves next to the dynamics of fish populations. Chapter 2 is a good illustration of the subtle blend of observations and theoretical analysis needed to obtain results in this field. The evidence that recruitment (annual numbers of new recruits joining a particular fish stock, say, at age 3) rises as a function of the population of mature stock to a maximum at a certain level of population *and then falls off* as the population increases beyond that level, is reviewed in some detail. A wide range of models of the population dynamics is also reviewed. These indicate that the extent of recruitment reduction at higher levels of the mature-fish population cannot be explained by any models which do not allow for a truly density-dependent mortality. Indeed, the combination of careful data analysis and dynamic modelling is shown to yield strong evidence that cod and related species may, essentially, control population levels by cannibalism.

Optimization theory is one of the favourite modern tools of applied mathematics which has also appealed strongly to biologists. This is because of the light which it can be expected to throw on the workings of the evolutionary process. Its use in this context is well exemplified by Chapter 3. Mathematical models are there used to assess the evolutionary advantages to gadoid fishes (cod, haddock, whiting, etc.) of different biological controls on the age of first maturation and on the proportion of total energy intake going into reproductive activity. Geographically separate stocks of the same species can show vastly different behaviour in these respects which are of importance to fisheries management. The optimization analysis of Chapter 3 makes a considerable addition to our understanding of the factors which may contribute to these differences, and of how they could change under different circumstances.

After the first three chapters, which are concerned primarily with the dynamics of natural populations, the book turns to the population dynamics of fisheries as such. (For background reading in this field see especially the classic book "On the Dynamics of Exploited Fish Populations" by R. J. H. Beverton and S. J. Holt. Her Majesty's Stationary Office, 1957.) Scientists advising governments on fishery management need above all to be able to make predictions on total numbers within a fish stock: total numbers which can be subdivided into the numbers in each year-class within the stock. These skills depend on obtaining good estimates of many variables, some of which like recruitment and *natural* mortality have already been mentioned. Another variable (indeed, one of special importance) is the *fishing* mortality. Statistical methods of considerable mathematical subtlety are needed to estimate it from the available data.

Chapter 4 discusses the statistical problem of estimating fishing mortality (rate of population loss by fishing as a function of time and of fish age) when the natural mortality together with the catch-at-age data (fish catch as a function of time and of fish age) are assumed known. Some rather special mathematical skills may be needed in problems like this which, though of great practical importance, are from the strict mathematical point of view "not well posed"! Chapter 4 demonstrates the essential non-uniqueness of the solution but shows how intelligent application of maximum-likelihood ideas can get round the difficulty. The chapter has further importance to fisheries science, however, because it presents such clear evidence for a remarkably high *sensitivity* of the inferred fishing mortalities *to errors* in the catch-at-age data.

Another route to fishing-mortality results is quantitatively explored in Chapter 5. This is the route that utilizes the results of large-scale "tagging" experiments (where fish are captured, tagged and released, and tagged fish are later identified in catches). The emphasis in the statistical analysis is again on making clear the "confidence limits" on the inferred mortality values (here, both fishing mortality and natural mortality). It is significant that a model taking into account diffusion (randomly increasing separations of the tagged fish released at a particular position) is included.

Appropriately, the book's first half, concerned with population dynamics, concludes with a partial account of the highly comprehensive type of mathematical model which, if adequate data existed, might simultaneously allow for all of the considerations mentioned above. A multispecies model might take into account the population dynamics of the phytoplankton, and of the herbivores, and of several links in the "food chain" to the fish species of commercial interest. It might take account, not only of fishing mortality, but also of the many components of natural mortality, including the complicated web of prey–predator relationships. Chapter 6 is concerned especially with this latter aspect of such a mathematical model, and with how a program should mathematically represent the dependence of natural mortality on the other important variables. Readers may find it interesting to speculate, in the light of all the information in Chapters 1 to 6, on when models of such a highly comprehensive type may become effectively usable.

All the work on population dynamics has as its principal end-product the determination of governmental and intergovernmental *restrictions on catch* (through quotas, mesh-size restrictions, etc.) which are needed to avoid over-fishing and to maintain fish stocks. On the other hand, within the limits of such catch restrictions, it has always been essential that the fish which are caught should be caught economically. The current high cost levels, especially of fuel and of crew, make this a matter of paramount importance.

Mathematical analysis has much to contribute to this area also, as is made clear in Chapters 7 and 8. Efficient methods of locating fish (often through the use of sonar devices) do much to minimize the time at sea needed to produce a given catch; such methods need careful quantitative interpretation. Trawl design is another important factor to which some complex analytical methods have been devoted. Chapter 7 is concerned with these problems, and with the methods of evaluating the data on performance trawls of different designs.

The sea, and its uses, are changing rapidly; many of the best fishing grounds are areas of exploration and exploitation for oil and natural gas, for sand and gravel, for cable-laying and so on. These newer uses do not prevent continued use of the areas as fishing grounds, provided that trawler skippers are kept supplied with the latest information on the location of any structures associated with those different uses (as well as of any wrecks, etc.). Accordingly, charts essential to fishermen need to be continuously updated. Here there has been an interesting challenge: to find ways of storing cartographic information in digital form in a computer "databank" which can be regularly updated and can yield graphic output as and when required, in the form wanted by fishing vessels. Chapter 8 describes a remarkable success story in this field.

The book ends with three chapters on aspects of fish behaviour significant in relation to fish catching. The general area of *animal locomotion* is one where substantial advances in quantitative analysis have been made in recent years. These are aimed, for example, at relating speed of locomotion to available muscle power, and also to the size of the animal. Somewhat similar advances have been made in such studies of terrestrial, aquatic and aerial locomotion (see the forthcoming book "Scale Effects in Animal Locomotion" (T. J. Pedley, ed.), Academic Press, 1977; for swimming and flying, see "Swimming and Flying in Nature" (T. Y. Wu, C. J. Brokaw and C. Brennen, eds.), Plenum Press, 1976; and also "Mathematical Biofluiddynamics" by J. Lighthill, *SIAM*, 1976.) For fisheries purposes, it is particularly important to relate swimming speeds to the power output of the two types of musculature that can be involved: the main "white" musculature usable only for relatively short bursts of fast swimming, and the modest volume of "red" musculature available for continuous swimming at much slower speeds.

A standard approach to the quantitative analysis of swimming by many fishes of commercial interest is the "elongated-body" (or "slender-body") theory of swimming described in Lighthill's book in both a "small-amplitude" and a "large-amplitude" version. Chapter 9 outlines that approach briefly, and then concentrates on the modification that may be needed, especially for fast-swimming fishes, such as tuna and other scombroid fishes with caudal fins of relatively high aspect-ratio. The hydrodynamics of such fins may be treated by methods pioneered in the aerodynamic field, while the hydrodynamics of the body may be analysed by elongated-body theory. Chapter 9 is particularly concerned with the means by which structural features of these fishes act to minimize the passive (but energy-wasting) yawing movements with which the anterior part of the body responds to those caudal-fin motions.

An interesting parallel with flying animals is indicated in the last chapter. It has long been known that the larger birds save energy by "soaring" motions in which, effectively, they extract energy from natural air movements. Chapter 10 gives evidence that migrating fishes, such as plaice, secure a similar saving of energy; effectively, they extract energy from natural water movements.

For fishes such as cod, the central problems of relating thrust both to power consumption and to the hydrodynamic resistance to be overcome are tackled in Chapter 11 by a combined experimental and theoretical approach. A well-equipped tank of 10 m diameter has been employed to obtain unusually accurate records of cod swimming movements under different conditions. The large-amplitude elongated-body theory is used to estimate the thrust developed by those swimming movements and to relate it to muscle power and hydrodynamic resistance, so that many of the basic questions of cod locomotion receive for the first time (even if only tentatively) some very plausible quantitative answers.

This book represents a joint effort, by the Marine Laboratory (Aberdeen) of the Department of Agriculture and Fisheries for Scotland and by the Institute of Mathematics and its Applications, to make a broad survey of "Fisheries Mathematics"; a survey which, from the outset, was planned to be equally of value to fisheries scientists concerned with modern quantitative methods of analysis and to mathematicians interested in studying in depth this absorbing field of application. The material was first presented in the course of a Conference held at the Marine Laboratory, Aberdeen, from November 24 to 26 1975; the two organizations concerned had, however, exercised great care in their choice of the speakers and the subject-matter so that a well-balanced written account of the field would emerge from the Conference.

Warm thanks are given to Mr. B. Parrish, the Director of the Laboratory, for presiding at the meeting, to Dr. J. H. Steele, the Deputy Director, for editing the Proceedings, and to many members of their staff for taking part, also to all the other participants, and to Miss Catherine Richards and the whole IMA staff for their efficient contributions to the organization of the meeting and the preparation of this book for the press. All those concerned with the book believe that its publication will act to strengthen still further the already strong links joining fisheries scientists and applied mathematicians.

JAMES LIGHTHILL

University of Cambridge

Contents

1. Plankton Patches in the Northern North Sea

J. H. STEELE AND E. W. HENDERSON

Marine Laboratory, Aberdeen

1. Introduction

This paper will be concerned with the pelagic ecosystem in the open sea; with nutrients, phytoplankton and herbivores; and with the effect of variations in the physical environment on these factors. The problems concern how to collect data, how to analyse them and how, conceptually or mathematically, to produce models of the process involved in the interactions of these factors. The basic problem, particularly in sampling, but hence in the other aspects, is on what time and space scales the data should be collected.

The critical scales, at our present level of knowledge and technique, appear to be the meso-scales of the order of kilometres and days (Platt and Denman, 1975), rather than very fine structure or yearly cycles over large ocean areas. Data collected, traditionally, at positions 10–100 km apart and at weekly or monthly intervals, with considerable averaging, may show certain accepted features such as the progressive changes in nutrients, chlorophyll and zooplankton dry weight during the spring outburst (Fig. 1). The problem is that the raw data show such great variance (Fig. 1) that it is often too easy to construct simulation models which lie within the broad limits of these data.

An alternative is to make more detailed and intensive studies over shorter periods, but these can contain as much variance as the larger surveys (Steele, 1974). The problem lies in the analysis of this variance, not only for statistical reasons, but also because, ecologically, the nature of this variance is of considerable importance to the animals themselves. Fish larvae are likely to feed successfully only in food concentrations well above the general average (Jones and Hall, 1973; Lasker, 1975). As Platt and Denman (1975) have pointed out, this problem involves the determination of the space and time scales of biological patchiness in the environment and the relation of these to similar physical distributions. Platt (1972) showed that, on scales less

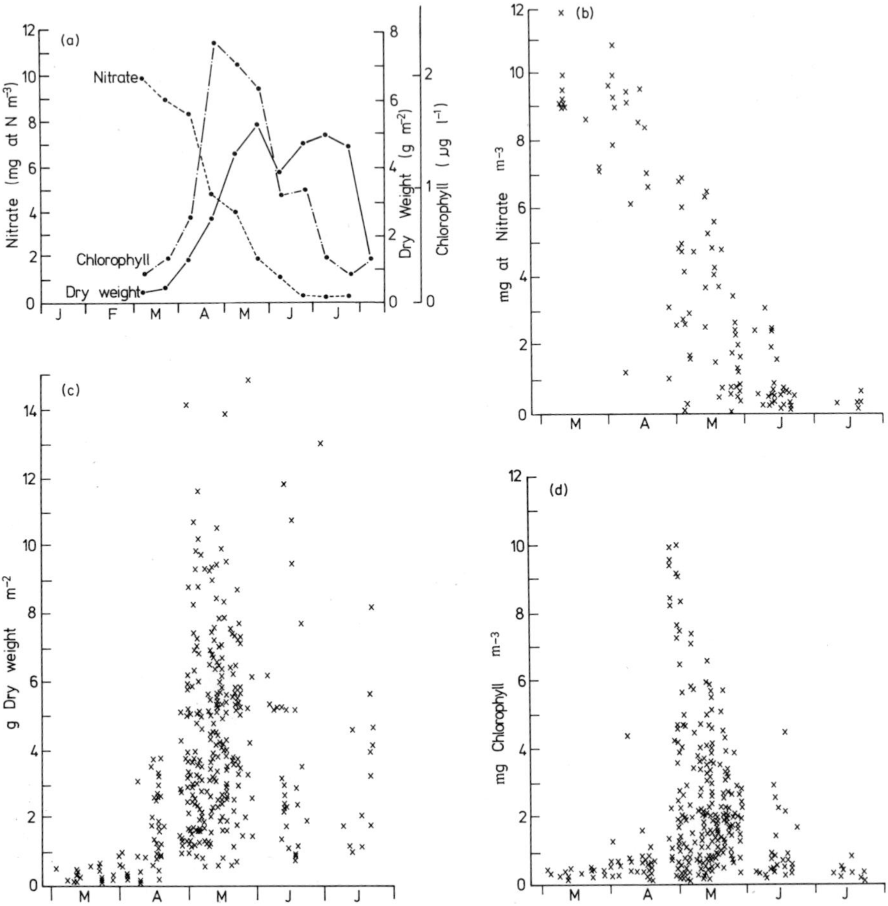

FIG. 1. (a) Average cycle of nitrate, chlorophyll and zooplankton dry weight in the centre of the northern North Sea, 1961–1970 (b) Raw data for nitrate. (c) Raw data for chlorophyll. (d) Raw data for zooplankton. (Adams and Martin, unpublished.)

than 100 m, the spectral distribution of chlorophyll was similar to that of the turbulence spectrum and so the former was effectively determined by the latter. Following Kierstead and Slobodkin (1953), it seems likely that for phytoplankton growth there is a critical scale of the order of a few kilometres (Steele, 1975) below which turbulent diffusion will mask the perturbing effects of phytoplankton growth. For this reason data collections in the North Sea in 1974 were intended to sample nutrients and phytoplankton for spectral analysis in the range 1–30 km.

2. Data Collection

On seven occasions (Table 1) between 4 and 29 April 1974 the ship steamed at 6 knots along a 240 km line (approximately 61°00′ N–59°00′ N; 0°30′ E).

TABLE 1. Runs during April 1974

Run	Date	Time (hours)	From	To
1	6	0001–1830	61°00′ N; 00°33′ E	59°01′ N; 00°30′ E
2	7–8	1600–1400	61°00′ N; 00°30′ E	59°21′ N; 00°31′ E
3	10–11	1230–1830	61°00′ N; 00°30′ E	58°48′ N; 00°29′ E
4	15–16	0600–0124	58°59′ N; 00°30′ E	60°58′ N; 00°30′ E
5	19	0600–2238	61°00′ N; 00°30′ E	59°02′ N; 00°30′ E
6	23–24	2010–1430	60°59′ N; 00°30′ E	59°02½′ N; 00°30′ E
7	29–30	1020–0600	61°01′ N; 00°30′ E	59°01′ N; 00°30′ E

Water intake from 3 m was sampled for temperature, salinity, nitrate and chlorophyll (Figs. 2 and 3), and the data read at 2·5 minute intervals (approximately 0·5 km). Temperature and salinity were recorded on a thermosalinograph; chlorophyll was obtained from *in vivo* fluorescence measurements calibrated at half-hour intervals by *in vitro* chlorophyll determinations;

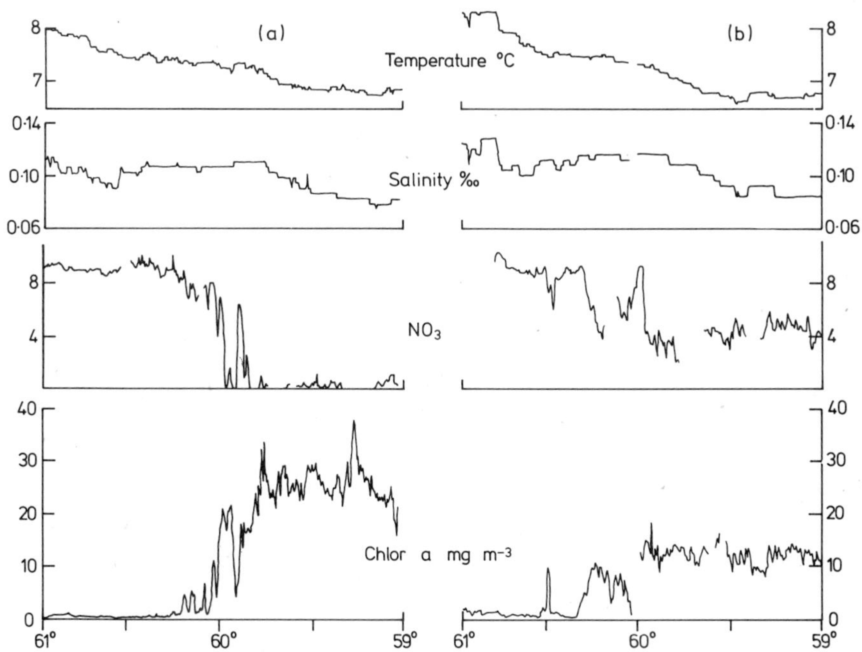

FIG. 2. Temperature, salinity, nitrate and chlorophyll from 3 m on (a) Run 1, (b) Run 3.

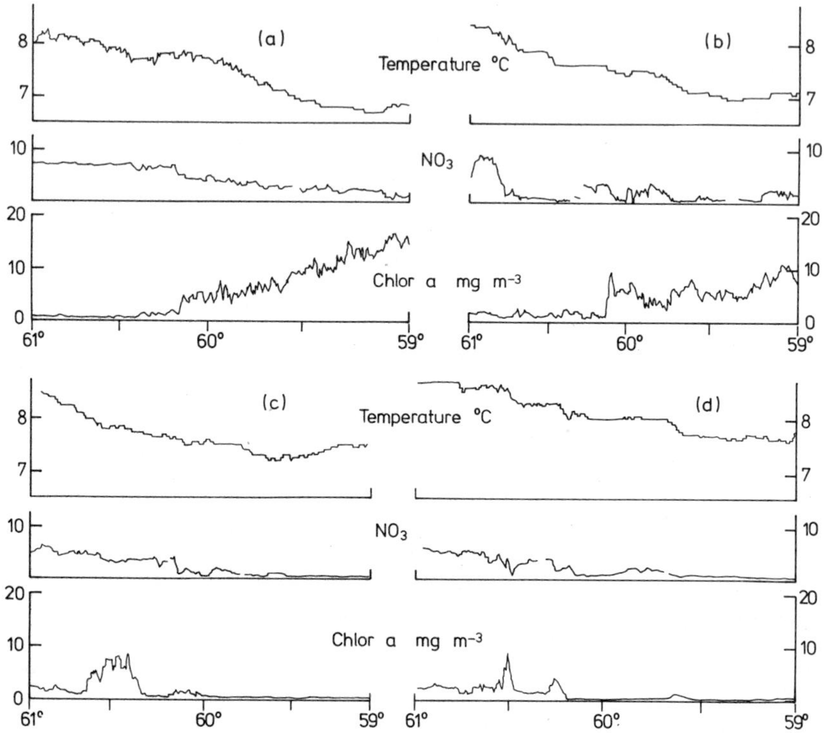

FIG. 3. Temperature, nitrate and chlorophyll from 3 m on (a) Run 4, (b) Run 5, (c) Run 6 and (d) Run 7, April 1974.

nitrate was measured on an Autoanalyser. During the passage of the water from intake to instruments in the laboratory some smoothing will occur. From experiments in similar pipes this will affect time intervals shorter than 1 minute. In the Autoanalyser, which takes 7·5 minutes to analyse a sample, further smoothing can occur, but this effect was estimated to be less than 1 minute. These factors combined should not affect the spectral analysis of readings at 2·5 minutes. Zooplankton were sampled by an oblique net haul from near bottom to surface (Fig. 4); the hauls taking 20 minutes in each half-hour. On the return trip to the other end of the section, stations were worked at 40 km intervals to give data on the vertical distributions of the same parameters. Although the sections took about 20 hours to run they will each be regarded as instantaneous spatial distributions.

This form of data acquisition gives good spatial cover at one depth. It does not provide detailed data in the vertical. The choice of which dimension

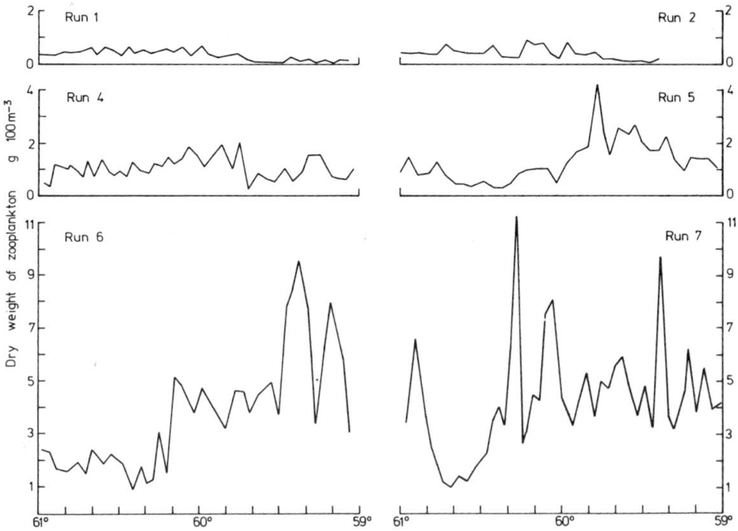

FIG. 4. Zooplankton dry weights from Runs 1–7 (250 μm mesh). (Run 3 is omitted since only part of the line was sampled.)

to work in most intensively is one problem inherent in the sampling design and is dependent on the magnitude of the ship operation. Another problem is that, in this survey, data on zooplankton could not be collected at the same intensity as for the other parameters.

The physical data show that the environment, as indicated by near surface temperature and salinity, is not uniform but has no regions of sudden change. Further, during this survey there were no marked changes in this pattern and this is confirmed by current meter data which show generally slow water movements. On the other hand, there are very marked changes in nitrate and chlorophyll especially in the earlier runs. It is of interest to note how the correlation (Fig. 5) changes from a high negative to a marginally significant positive value during the course of the survey. This kind of change is one reason why data more widely spaced in time and location are unlikely to yield to straightforward correlation techniques. A partial explanation for these changes in correlation can be derived from the zooplankton data (Fig. 6) which have been averaged over 60 km blocks from the runs and compared with the average chlorophyll values. Initially, in all blocks zooplankton were very low so that an increase in phytoplankton would correspond to an uptake of nitrate and, with no grazing on the phytoplankton,

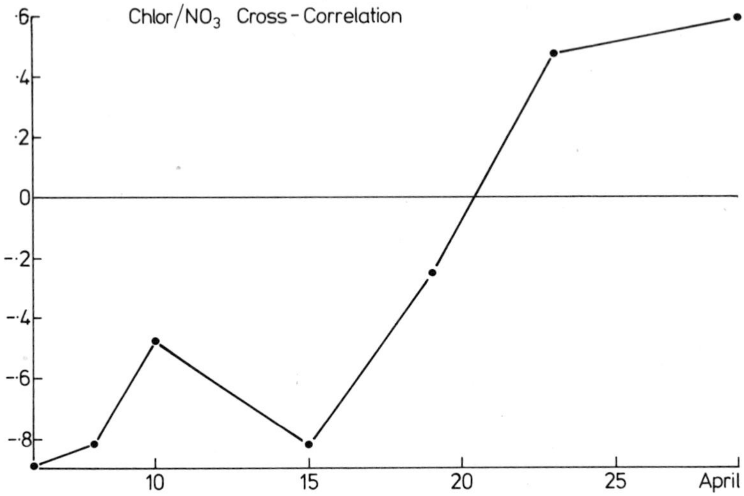

FIG. 5. Correlation, using the raw data from Runs 1–7, between chlorophyll and nitrate.

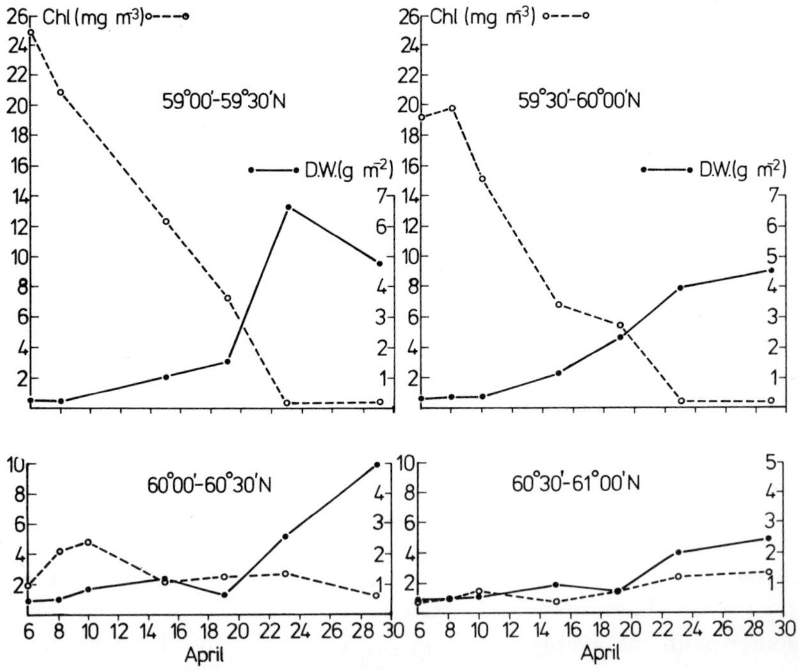

FIG. 6. Average chlorophyll and zooplankton dry weights within sections of each run.

the increase in chlorophyll concentration will then mirror the nitrate decrease. As the zooplankton develop this simple relation breaks down as a result of the effects of grazing.

The close similarity in the inverse changes in the two southern blocks (Fig. 6) is some measure of agreement over relatively short spatial separations. The complete difference between the form of these changes and the increase together of chlorophyll and zooplankton in the northern block is a further example of the lack of simple general statistical relations arising from such interactions.

The divergence between the north and south of the line arises from the differing density distributions in these two regions (Fig. 7), at the start of the survey. The relatively strong gradients near the surface in the southern area produced conditions suitable for the rapid growth of phytoplankton. One feature, apparent in the southern area, is the breakdown of this thermocline due to strong winds between 8 and 11 April. These followed a period of very

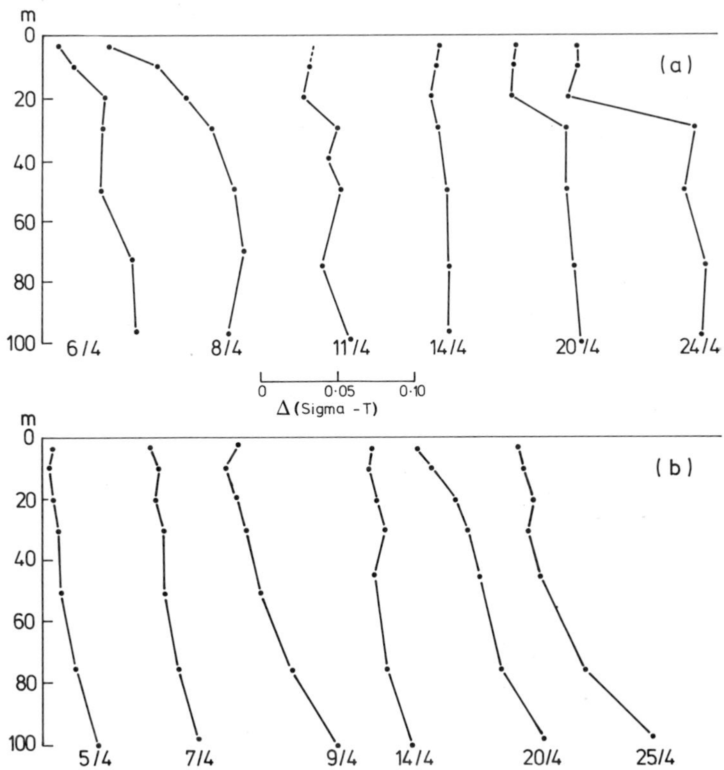

FIG. 7. Density distribution at (a) 59°40′ N and (b) 60°40′ N.

quiet weather which, presumably, was a major factor in establishing the shallow thermocline. A consequence of the breakdown is the redistribution vertically of nitrate and chlorophyll (Fig. 8). In fact, these parameters, rather than density, provide the best indication of the effects of the increased vertical mixing. This illustrates one problem in simulation since, theoretically, the physical data should be used to determine the redistribution of chemical and biological factors, rather than *vice versa*.

3. Horizontal Variability

These variations in depth distribution with latitude will affect the study of horizontal changes in the physical and biological parameters. Thus studies of horizontal variations need to be related to vertical changes although the following analysis considers them separately.

The general negative correlation between nitrate and chlorophyll over the

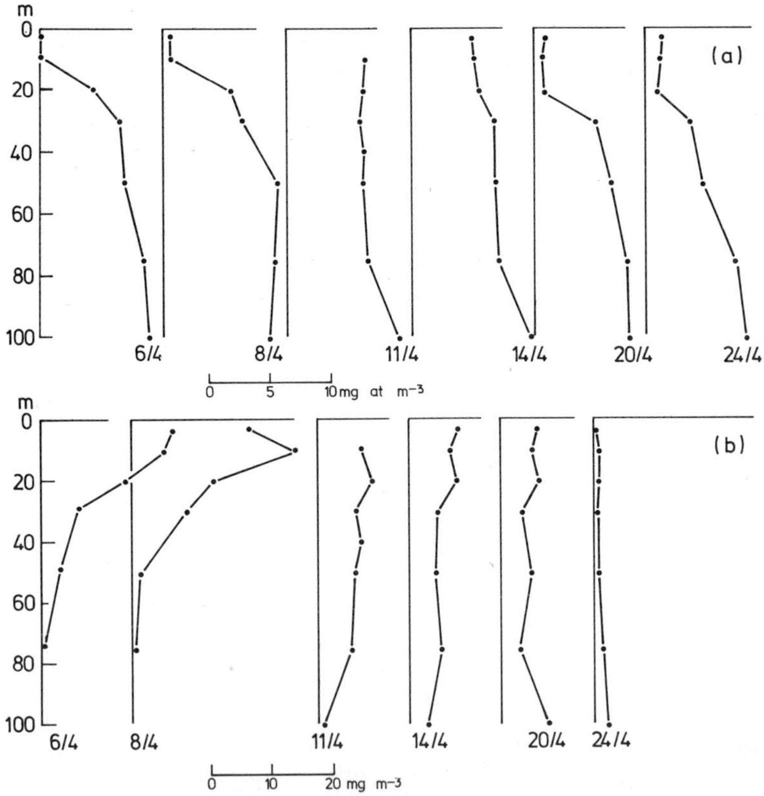

FIG. 8. Vertical distribution of (a) nitrate and (b) chlorophyll at 59°40′ N.

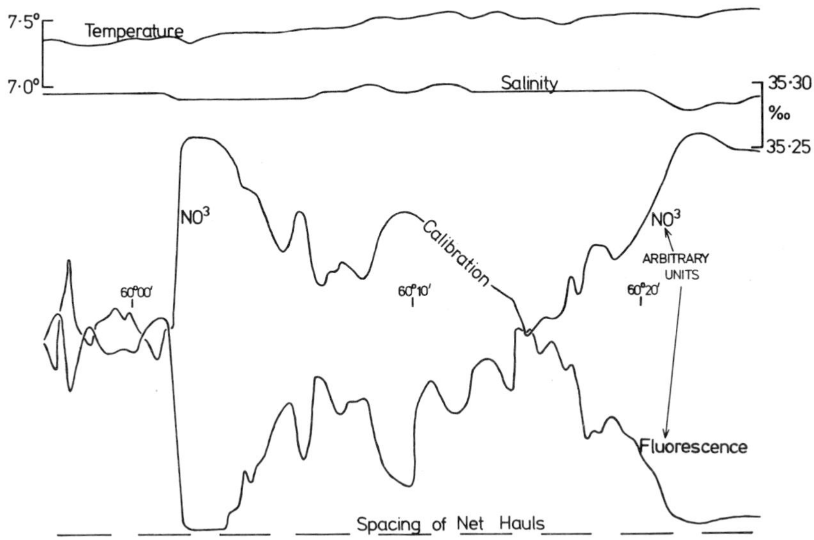

FIG. 9. Details of data from part of Run 1.

whole of Run 1 is still found at the smallest scales observed (Fig. 9). This illustrates the need to look at variability on all scales. Figure 9 shows the lack of any obvious correspondence with near surface temperature and salinity changes and also indicates the larger scale of the zooplankton samples. For the spectral analysis, temperature and chlorophyll have been used as representative of the physical and biological parts of the system. Analyses for the whole records (Fig. 10) are typical of such data (e.g. Platt and Denman, 1975). The generally similar slope of the responses implies, as expected, that the variability of the physical environment, associated with the spectrum of turbulence, plays a major role in determining the spectrum of the biological components (Platt, 1972). There are, however, differences in detail in the slopes and, for the chlorophyll, in the magnitude of the variance. If, as a hypothesis, one assumes that near surface temperature is a good index of the input of variability to the biological system, and chlorophyll of the output, then, regarding the physical/biological interactions as a "black box", the ratio of spectral energies can be used as an index of transfer within this system. It is necessary to take account of the changes in average value of the chlorophyll for each run. For biological data the variance changes with the mean (Cassie, 1963). Tests with these data showed that simple proportionality of variance to mean gave the least overall change from Run 1 to Run 7, so, for these data, the variances for each run

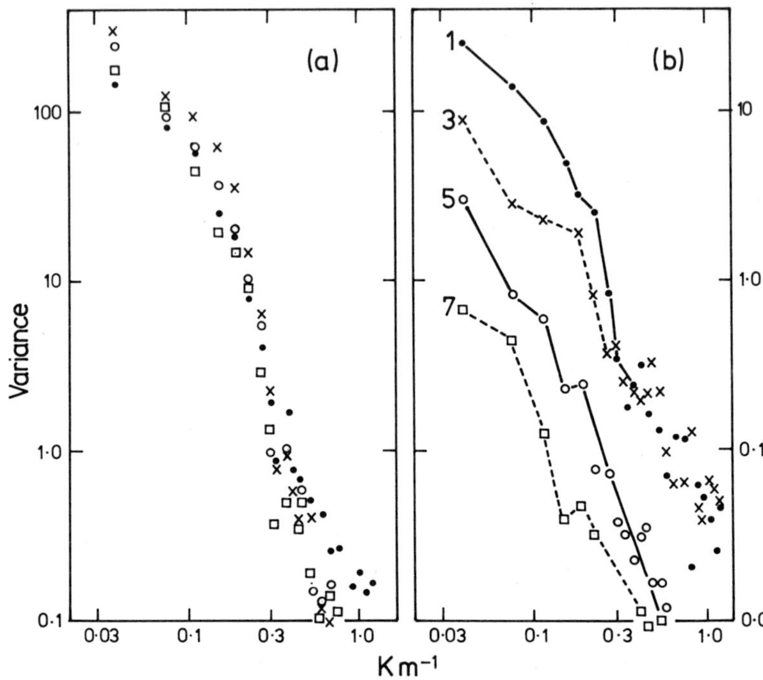

FIG. 10. Spectral analysis of (a) temperature and (b) chlorophyll data from alternate runs.

have been divided by the mean for that run before calculating the spectral ratio (Fig. 11). There are two main features of these results. Firstly, the general nature of the change with wavenumber is remarkably similar over the six runs used here (Run 2 has not been used since there was a break of several hours in the middle of the run). Secondly, the absolute values of the spectra show a marked decrease between Run 3 and Run 4 with a recovery thereafter. This change results from changes in the output (chlorophyll) component. Between Runs 3 and 4 there were the very strong winds mentioned earlier and these would be expected to result in a marked increase in horizontal as well as vertical mixing. Although the near-surface temperatures on the runs do not show this, data from a thermistor moored at 28 m depth in the southern half of the section indicate the smoothing resulting from the increased mixing (Fig. 12). A moored thermistor gives neither pure temporal nor spatial data because of tidal flow past it. Thus Fig. 12 can only provide general indications of the probability that variability of near surface chlorophyll data is dependent on subsurface physical structure.

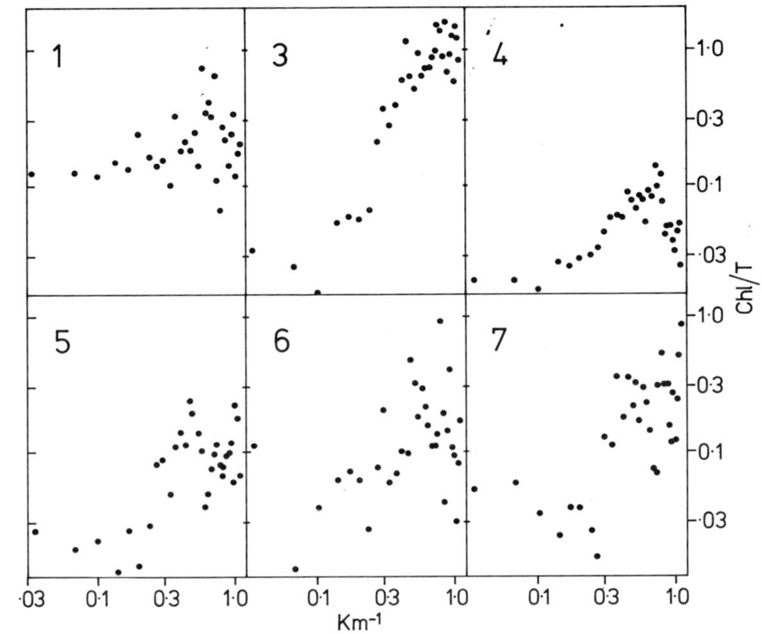

FIG. 11. Spectral ratios (except Run 2 which had a time gap in the middle) with chlorophyll data corrected for mean value (see text).

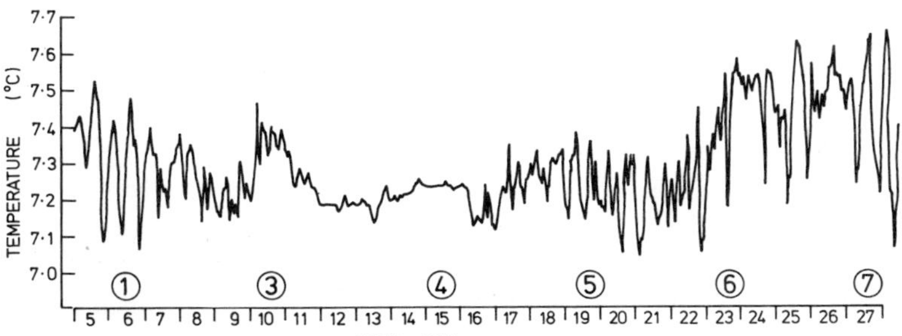

FIG. 12. Hourly means of thermistor data from 60° N at 23 m depth with time of runs indicated.

4. Nitrate–Temperature Relations

A similar treatment can be made of the nitrate data. In this case there are gaps in the data, where calibrations were carried out, amounting to about 3% of the total data. Linear interpolation has been used to fill these gaps. The nitrate data were analysed spectrally and then divided by the temperature spectra to produce nitrate/temperature ratios (N/T) (Fig. 13) comparable

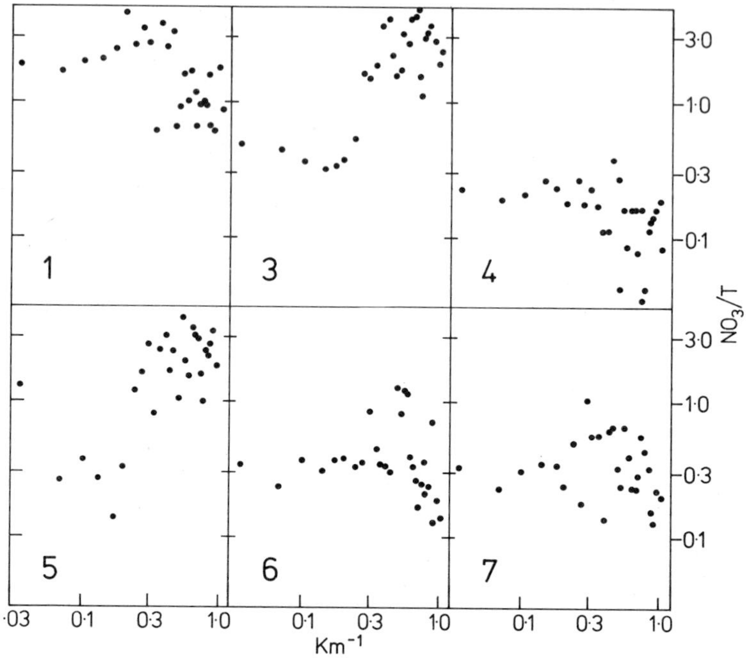

FIG. 13. Spectral ratios for nitrate/temperature data.

to those for chlorophyll/temperature (Chl/T) except that no correction was made for mean value of N. These N/T ratios do not show any regular trend with wavenumber. The only feature in common with the Chl/T ratios is the reduction in absolute value during Run 4 which would, again, indicate the effects of the gale. It is possible that the passage of the water through the Autoanalyser may have smoothed some of the high frequency but the observed delay of 7·5 minutes would, at most, have affected scales of 0·67 km^{-1} and the differences between N/T and Chl/T appear to occur down to lower wavenumbers.

5. Space–Wavenumber Distributions

It is apparent from the data (Figs. 3 and 9) that there are likely to be variations in the nature of the spectra for different geographical positions on each line. To investigate this each record was transformed to a sequence having zero mean value. Each sequence was then split into overlapping segments 64 time units long; the distance between the centres of these segments being 32 time units. Raw spectral estimates were calculated using a Fast Fourier Transform and the results were smoothed by frequency averaging over three adjacent estimates. (The results in Fig. 10 were, in fact, obtained by segment averaging over all the segments in the sequence.)

The data, presented in this way, show that many of the features observed in the spatial average (Fig. 10) arise from particular parts of the geographical section. For example, in Fig. 14, the distribution in Run 3 has high values at low wavenumbers concentrated in the boundary region between the high and low chlorophylls. After the gale (Run 4, Fig. 14) this became an area of minimum variance. The occurrence of abrupt chlorophyll changes in this region (Fig. 9) suggests that subsurface "fronts" existed undetected by surface temperature, and that these were smoothed out by the effects of strong winds.

6. Modelling of Horizontal Structure

Attempts to simulate the spatial heterogeneity are at a more primitive level than for depth/time models. The original developments considered only the interactions of phytoplankton growth with turbulent diffusion (Kierstead and Slobodkin, 1953). Combining these with data on turbulent diffusion rates as a function of scale (Okubo, 1974) led to estimates of a critical scale for a phytoplankton patch of the order of a few kilometres (e.g. Steele, 1975). Below this scale a patch would be damped out, above it the patch would grow until other factors such as nutrient limitation or grazing came into play. On this basis, below a wavenumber between, say, 0·1 and 1·0 there should be a greater response than above it. The responses observed (Fig. 10) are opposite to this.

As a first step to greater complexity, the Lotka–Volterra (prey–predator) equations have been combined with horizontal diffusion terms (see Appendix). This extra complexity has two effects. Firstly, there is no longer a critical length for any scale or magnitude of initial perturbation (Murray, 1975). Secondly, as can be expected from the introduction of non-linear interactions, there will be transfers of variance between wavenumbers (Steele, 1974). The investigation of this effect requires a simple simulation but, because of the damping effect of diffusion at all scales, it is necessary to introduce perturbations, stochastically, throughout the run of the simulation. Technically this can be done in several ways and most simply by an addition of random numbers to one or both of the phytoplankton and herbivore equations. The empirical problem is to provide a physical basis for such an addition. External factors such as variations in incident radiation (required for photosynthesis of organic matter by the phytoplankton), or internal waves, could be used as possible mechanisms. The generally unsatisfactory

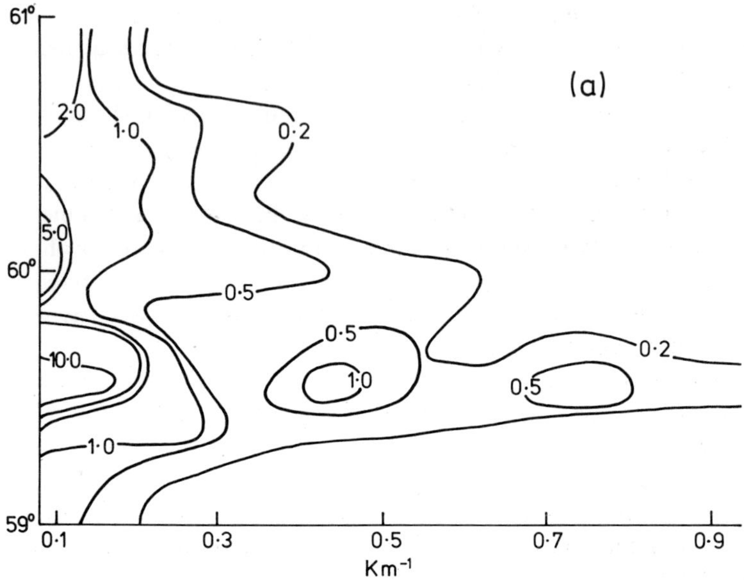

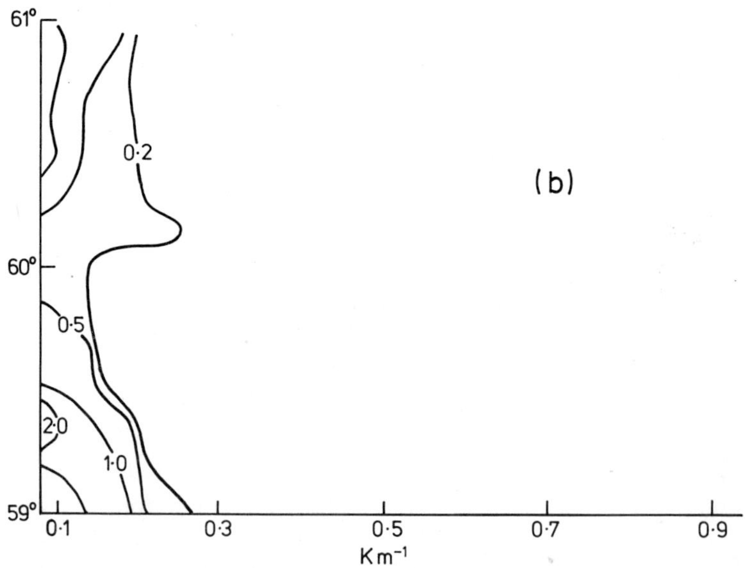

FIG. 14. Plots of spectral distributions of chlorophyll at different positions along the section for (a) Run 3 and (b) Run 4.

nature of this assumption is discussed in the last section. Since the scale of the temperature spectrum approximates to k^{-2}, the random numbers were scaled by a factor n^{-1} (see Appendix).

The output from such a simulation with roughly "realistic" constants can be divided by the scaling factor, n^{-1}, to give a spectral ratio comparable in principle to that used in the data analysis (Fig. 15). The result, however, is in the wrong direction when compared with the data. Other runs with different inputs and constants have not produced outputs differing in character from the one shown here.

There are three probable inadequacies in this comparison of simple theory and observation.

(*i*) The use of surface temperature as an "input" function may not represent those factors in the physical environment most critically affecting the biological system. The subsurface thermistor data indicate that this is highly probable.

(*ii*) The Lotka–Volterra equations are accepted as an inadequate representation of even the simplest ecosystem. More complicated biological interactions could, possibly, introduce oscillations at particular wavenumbers.

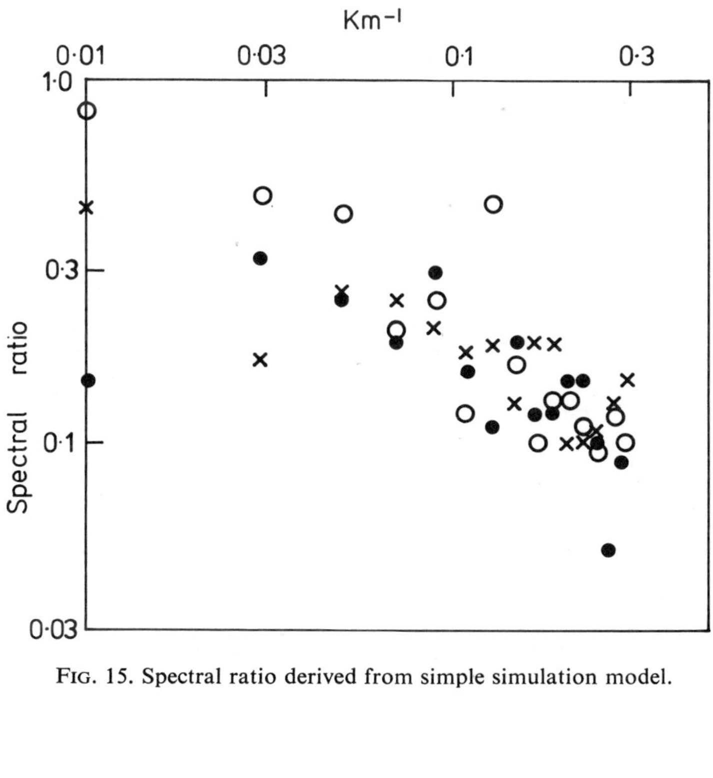

FIG. 15. Spectral ratio derived from simple simulation model.

However, one has to guard against looking for a mathematical form which will do this, thus forcing the biology into this form.

(*iii*) The main inadequacy is possibly in the representation of the mixing processes by simple diffusion terms. That this is inadequate is shown already by the need to add in stochastic variations. This is unlikely to be remedied by minor changes in input of variability. The concept of "horizontal turbulent diffusion" is a simplification of several processes such as horizontal velocity shear, thermocline deepening and internal waves. It is likely that a direct input of these factors to biological models will be necessary. This will require at least a two (X/Z) dimensional model and so will necessarily include the interactions in these two dimensions.

7. Conclusions

The comparison of the earlier data collection over large areas and time scales with the detailed runs made in 1974 illustrates the general problems of great variability in ecological data. Taken together they show the wide range of scales on which this variability occurs. It is possible to see some regularities in these data by comparing chlorophyll and zooplankton in spatially adjacent blocks of data or correlating, say, chlorophyll and nitrate for particular runs, but such relations do not extend far in time or space. On the other hand, spectral analysis of the variance does show regularities which for certain data extend over the whole sequence of runs. This corroborates the general conclusion of Platt and Denman (1975) that this method of presentation may provide a more useful "picture" of the world than straightforward space–time representation. It is also possible to combine the two approaches by relating spectral and spatial distribution of variance; and this may provide further insight into the causes of such variance in terms of events in the physical environment.

The simple "prey–predator" simulation of variance as a function of wavenumber shows that it is possible to develop a theoretical approach which permits direct comparison of theory and data. The fact that these can be seen not to match in this case is an indication of the advantages of this form of data analysis compared with a space–time presentation where the variability may accommodate theory too easily.

The trend for increasing spectral ratio (chlorophyll/temperature) with wavenumber occurred over the whole seven runs. This covered a period of time from an initial condition when zooplankton were negligible and phytoplankton growth dominated to a later period when zooplankton were probably a major factor limiting phytoplankton growth. Thus an explanation of this ratio in terms of phytoplankton growth only, or in terms of phytoplankton–zooplankton interactions only, would seem to be inadequate. Data collection and simulation will need to include more complicated aspects of the dynamics of the physical environment.

Acknowledgements

We thank Mr. J. Pirie, Mr. I. E. Baird and Mr. D. V. P. Conway for assistance with nutrient, chlorophyll and dry weight data collection and analysis.

References

CASSIE, R. M. (1963). Microdistribution of plankton. *Oceanogr. Mar. Biol. Ann. Rev.* **1**, 223–252.

JONES, R. and HALL, W. B. (1973). A simulation model for studying the population dynamics of some fish species. *In* "The Mathematical Theory of the Dynamics of Biological Populations" (M. S. Bartlett and R. W. Hiorns, eds.), pp. 35–59. Academic Press, London and New York.

KIERSTEAD, H. and SLOBODKIN, L. B. (1953). The size of water masses containing plankton blooms. *J. Mar. Res.* **12**, 141–147.

LASKER, R. (1975). Field criteria for survival of anchovy larvae: the relation between inshore chlorophyll maximum layers and successful first feeding. *Fishery Bulletin* **73**, 453–462.

MURRAY, J. D. (1975). Non-existence of wave solutions for the class of reaction–diffusion equations given by the Volterra interacting-population equations with diffusion. *J. Theor. Biol.* **52**, 459–469.

OKUBO, A. (1974). Diffusion-induced instability in model ecosystems: another possible explanation of patchiness. *Technical Report* 86. Chesapeake Bay Institute, The Johns Hopkins University.

PLATT, T. (1972). "The Feasibility of Mapping the Chlorophyll Distribution in the Gulf of St. Lawrence." Fish. Res. Board Can., Technical Report 332, 8pp. (20 figs.).

PLATT, T. and DENMAN, K. L. (1975). Spectral analysis in ecology. *Ann. Review of Ecology and Systematics* **6**, 189–210.

STEELE, J. H. (1974). "The Structure of Marine Ecosystems." Harvard University Press, Cambridge, Mass. 125pp.

STEELE, J. H. (1975). Biological modelling. *In* "Modelling of Marine Systems" (Jacques C. J. Nihoul, ed.), pp. 207–216. Elsevier Oceanography Series, Amsterdam, Oxford and New York.

Appendix

Let

$$P = \text{phytoplankton carbon (C mg m}^{-3}\text{)},$$

$$H = \text{herbivorous zooplankton carbon (C mg m}^{-3}\text{)}.$$

Then let the interaction of these populations be

$$\frac{\partial P}{\partial t} = aP - bPH + \frac{\partial}{\partial x} k \frac{\partial P}{\partial x}, \tag{1}$$

$$\frac{\partial H}{\partial t} = cPH - dH + \frac{\partial}{\partial x} k \frac{\partial H}{\partial x}, \tag{2}$$

where

a = phytoplankton growth rate,

b = grazing rate of a unit of H on P,

c = resultant growth rate of $H (c < b)$,

d = predation rate on H,

k = scale-dependent diffusion coefficient.

In the region $0 \leqslant x \leqslant L$ let

$$P = \sum_0^\infty A_n \cos \lambda n x, \quad \lambda = 2\pi/L,$$

$$H = \sum_0^\infty B_n \cos \lambda n x.$$

This assumes there is no net flux of P or H through $x = 0$, $x = L$. Then

$$\frac{dA_n}{dt} = A_n(a - \lambda^2 n^2 k) - bC_n \tag{3}$$

$$\frac{dB_n}{dt} = cC_n - B_n(d + \lambda^2 n^2 k), \tag{4}$$

where

$$C_n = \tfrac{1}{2}\left(\sum_{r+s=n} A_r B_s + \sum_{|r-s|=n} A_r B_s \right).$$

To introduce perturbations into the system add the term $\sum_{n=1}^\infty \alpha_n \cos \lambda_n x$ to the right-hand side of (1) where α_n varies randomly with time, within limits dependent on n.

The underlying excessively simple assumption is that some physical factors in the environment cause random variations in the growth rate of P. It is possible, although unsatisfactory, to think of these in terms of, say, incident radiation or internal waves. This will result in the addition of α_n to the right-hand side of (3).

The advantage of these very simplified equations is that their numerical solution can be worked out completely in terms of the Fourier coefficients. For this purpose only the first 30 coefficients in the series were used. The effect of this closure of the series has not been studied but from the marked decrease of A_n and B_n with increasing n, this should not affect the general nature of the results.

To parameterize the scale effect on the diffusion coefficient

$$k = k^*.n^{-1}$$

where k^* is a constant. This is an approximation to observed results (Okubo, 1974).

The values for α_n were taken from a random number generator and scaled to give

$$-n^{-1} \leqslant \alpha_n \leqslant n^{-1}, \quad 1 \leqslant n \leqslant 30.$$

Thus their variance should decrease as n^{-2}, in rough correspondence with the observed temperature data.

2. Development of a Model of Stock and Recruitment

D. H. CUSHING AND J. W. HORWOOD

Fisheries Laboratory, Lowestoft

1. Introduction

There are three current models of stabilization processes in fish populations. The simplest is the first one of Beverton and Holt (1957) which states that between hatching and recruitment there are two components of mortality, density independent and density dependent; then

$$R = \frac{1}{\alpha + \beta/P},$$

where R is recruitment, P is stock, α and β are constants that represent the two forms of mortality. In this model recruitment is asymptotic to stock.

The second slightly more complex formulation is that of Ricker (1958)

$$R = AP\exp(-BP),$$

in which the density-dependent mortality rate, B, is generated by the positive aggregation of predators at an early stage in the life history; A is the density-independent mortality rate. In this model recruitment may form a dome in stock.

The third model is that of Ricker and Foerster (1948) in which the death rate of larvae depends upon the time to grow through the attack field of a constant predator. If the food is abundant, the time is short and *vice versa*, so this mortality rate is density dependent. Beverton and Holt used this argument in their second formulation to reach the same equation as Ricker's but by a different route.

Garrod and Jones (1974) have described the dependence of recruitment in numbers on parent stock as eggs in the Arcto-Norwegian stock of cod and a pronounced dome in recruitment is shown (Fig. 1). The error lines shown are those to the curve fitted by least squares (Cushing and Harris, 1973), so the dome is real. Such a dome was originally noticed by Herrington

UNIV. COLL. N.W. BANGOR LIBRARY

(1948) in the stock of haddock on Georges Bank off the eastern seaboard of the United States. Very recently its existence has been confirmed by Sahrhage and Wagner (in press) on the observations on the North Sea haddock. The stock of cod in the Gulf of St. Lawrence exhibits the same feature (Paloheimo and Kohler, 1967). A reasonable generalization is that the stock/recruitment relationship in the cod-like fishes appears to be dome-like. Some stocks of Pacific salmon may exhibit the same character.

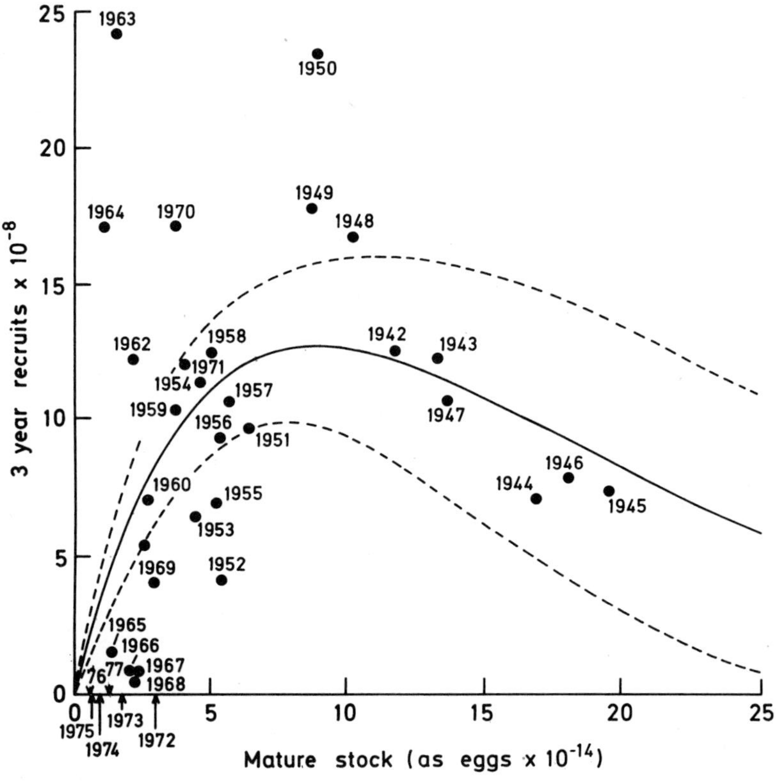

FIG. 1. The dependence of recruitment, in numbers, on stock (as eggs) in the Arcto-Norwegian cod population (Garrod and Jones, 1974); the dotted lines show the error curves of the least-square fit, i.e. there is one chance in 20 that the true curve lies outside the error lines.

Harris (1974) drew an important distinction between density dependence and stock dependence. The former can occur at any time during the life history as Beverton and Holt postulated, but the latter must take place early enough for the differences in stock from year-class to year-class to exert their influence. Only stock-dependent effects can generate the relationship observed between recruitment and parent stock. Ricker's formulation makes this point explicit, but the two Beverton and Holt models can only be stock dependent if the processes occur early enough in the lives of larval fish.

The Ricker–Foerster thesis implies that growth is density dependent. There is some evidence that in immature life growth is density dependent (Pacific halibut, Southward (1967); small gadoids, Raitt (1968); North Sea herring, Iles (1968)). In adult life, density dependence in growth has been difficult to detect; indeed, there is no published record of density-dependent growth amongst the older adult fish. This probably rules out the control

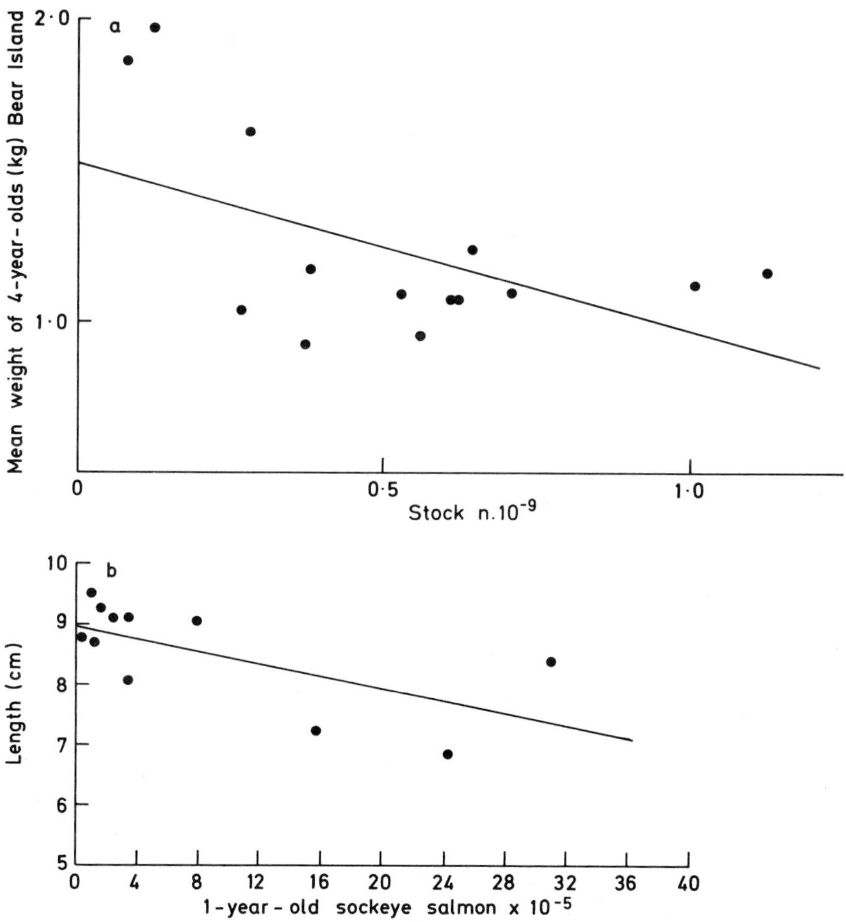

FIG. 2. (a) The inverse relationship between the weight of 4-year-old immature cod at Bear Island (from English landings) and numbers in the year-class at 4 years of age. (b) The inverse relationship between the length of 1-year-old sockeye salmon and their numbers in Cultus Lake (Foerster, 1968).

of recruitment processes by fecundity in fishes. However, Lett *et al.* (1975) have shown that the growth rate of cod between the ages of 5 and 8 in the Gulf of St. Lawrence is inversely related to biomass; the dependence on density has not yet been shown. For the Arcto-Norwegian stock of cod, Rollefsen (1954) suggested that density-dependent growth could be detected by comparing the mean length and stock numbers of the age groups 8–13 between 1932 and 1952; in fact the relationship is not quite significant at the 5% level, which illustrates the difficulty of establishing the existence of density-dependent growth even in a large fish. Figure 2(a) shows the inverse relationship of mean weight of 4-year-old cod from English landings off Bear Island on the stock in numbers of each year-class as at 4 years of age from the virtual population analysis on the total Arcto-Norwegian stock; an implicit assumption is that the numbers in the whole stock at 4 years of age represent those off Bear Island.

The 4-year-old fish which live in the Bear Island area are mainly immature; the figure does not pretend to describe the relationship properly, but merely indicates the existence of density-dependent growth. Saetersdal and Cadima (1960) made the same point when they showed clear relationships between growth constants and population density for young and old cod of 3 year-classes. Figure 2(b) shows an inverse regression of mean length of 1-year-old sockeye salmon in Cultus Lake (British Columbia) on numbers (Foerster, 1968) which again merely indicates the existence of density-dependent growth.

Each of the three models quoted above is descriptive in the sense that the parameters can only be extracted by fitting observations of recruitment to those of parent stock. To advance we need to establish the parameters independently. In the model to be described, the growth rates of larvae, juvenile and adult fishes are so derived, but the estimated mortality rates are taken from Garrod and Jones' figure. The density-independent mortality rate is the slope of the line at very low stock values and the density-dependent mortality rate was chosen such that the resulting stock and recruitment curve gave a good visual fit to that of Garrod and Jones up to middle stock values, i.e. lower than those at which the dome is formed.

2. Description of the Model

The model describes the growth and mortality of fishes from hatching to adult life.

(a) Mortality rate; Hatching to 3 years of age

We have used Beverton and Holt's formulation:

$$\frac{dN}{dt} = -(M_1 + M_2 N) N, \qquad (1)$$

where N is the number of larvae, M_1 is the density-independent rate of mortality and M_2 is the density dependent rate of mortality. M_1 and M_2 were estimated directly from Fig. 1 for stock values of 1–7·5 eggs $\times 10^{-14}$; $M_1 = 4{\cdot}0$ per year; $M_2 = 1{\cdot}5 \times 10^{-14}$ per year.

A second possibility is:

$$\frac{dN}{dt} = -M_1 N - \left(\frac{M_3}{W_t}\right) N^2 \tag{2}$$

where W_t is the weight of fish at age t. Here we assume that M_3 is constant, but (M_3/W_t) is a rough approximation to the Ricker–Foerster thesis; in fact, $M_3/W_t = M_2$ if there is no density-dependent growth.

A third possibility is the use of a stock-dependent component of mortality:

$$\frac{dN}{dt} = -(M_1 + M_2 N_0) N, \tag{3}$$

where N_0 is the initial number dependent on the size of the parent stock.

(*b*) *Mortality rate;* 3 *years of age onwards*

$$\frac{dN}{dt} = -(F+M)t, \tag{4}$$

where F is the fishing mortality, which increases to a maximum at 5 years of age and $M = 0{\cdot}2$; Garrod and Jones (1973) used $M = 0{\cdot}3$. There is little difference in the model between the two values. The natural mortality is assumed constant with age, which is sufficient for our present purpose.

(*c*) *Growth rate*

The growth rate of larval cod is not at all well known. Dannevig (1889) grew cod larvae (from a southern Norwegian fjord) to beyond metamorphosis from a single spawning, and the growth rate in length can be established. From larvae caught during a series of cruises in the Southern Bight of the North Sea (Harding *et al.*, in press), a length–weight ratio was established. Applying this to Dannevig's larvae from hatching to metamorphosis, a period of 70 days, the cod larvae grew at 10·35% per day; from 70 to 168 days, they grew in weight at 2·4% per day. To reach the observed weight at 3 years of age from 168 days, they would have to grow in weight at 0·71% per day; the weight at 3 years of age was taken from the Report of the ICES Northeast Arctic Working Group (1967). It might be argued that the growth rate of larvae in a southern Norwegian fjord does not represent that of Arctic cod larvae; Jones (1973) has found that the growth rate of North Sea haddock was about 10% per day and we might consider such values as representing larval gadoid growth rates.

Growth in weight from 3 years onwards was calculated with the von Bertalanffy equation:

$$W_t = W_\infty[1 - \exp(-Kt)]^3. \qquad (5)$$

Given K from the Report of the ICES Northeast Arctic Working Group (1967) and the weight at 3 years of age, the weight at infinite age, W_∞, is calculated. Then the weight at any age can be calculated, particularly that at age 8, the age of recruitment to the adult mature stock.

As noted in the introduction to this paper, the growth of fishes is probably density dependent, particularly amongst the immature age groups. The density dependence may decrease with age, with the result that it is difficult to detect in the fully recruited age groups. In the model, density-dependent growth was described as follows. If no competition for food existed, the fish would all be the same size at age 3, independently of N_0 the original stock in eggs. Competition was described by a parameter, S, such that the daily weight increment was modified as follows:

$$FN = [1{\cdot}0 - S(N_0 - \bar{N})], \qquad (6)$$

where FN is the coefficient that modifies the increment under competition; and $\bar{N}$ is a typical stock size at which the average weight was experienced. If $S = 0$ or if $N_0 = \bar{N}$, there will be no competition and the weights found will be the average weights experienced. If $S > 0$, the weight of fish at age 3 (or 8) will become progressively less as N_0 reaches a maximum.

The increment of growth with no competition for food is given by

$$\Delta W = W_1 C,$$

where W_1 is the initial weight and C is the average increment per day (e.g. 10·35% per day in larval life). When there is competition,

$$\Delta W = W_1 C(FN).$$

The model starts with a given stock number, as eggs, and recruitment is generated with various combinations of growth and mortality. In this way, a stock and recruitment curve can be described independently of observations of the two variables. It is true that our estimates of M_1 and M_2 are taken from the best fitted curve to the observations. However, the values are taken at low stock, whereas we try to describe the conditions for a dome in the curve of recruitment on parent stock, which is found at high stock.

There are three observations on the stock-recruitment curve of the Arcto-Norwegian cod. First, there is a pronounced dome in the relation between numbers of recruits and parent stock in eggs. Secondly, there is also a pronounced dome in the relation between recruits, as eggs, and stock as eggs (Garrod and Jones, 1974). Thirdly, growth in the Arcto-Norwegian cod is density dependent at least amongst immature fish. The model is manipulated to match the three observations.

3. Results

The model is used to test different combinations of growth and mortality. In the first combination, growth and mortality are as observed, but they are independent of each other, i.e. growth is not density dependent. Figure 3(a)

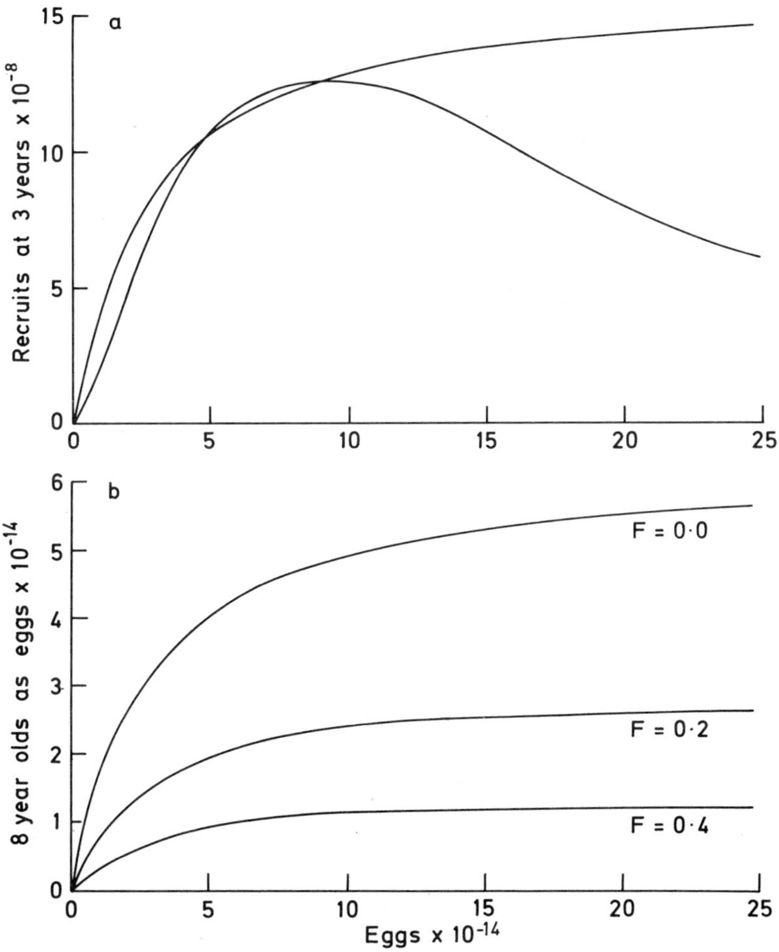

FIG. 3. (a) The asymptotic dependence of recruitment, in numbers, on stock (as eggs), with growth and mortality independent of each other; the dome is that in Fig. 1. (b) The dependence of recruitment, as eggs, on stock (as eggs), with growth and mortality independent of each other, at different levels of fishing mortality.

shows the asymptotic dependence of numbers of recruits on stock as eggs as compared with the observed dome (from Fig. 1). Figure 3(b) shows a similar relationship for recruits (as eggs) on stock (as eggs). There is no density-dependent growth so none of the observational requirements is satisfied. Hence the independence of growth and mortality is perhaps unlikely.

There are two ways in which growth and mortality may be linked. In the first, growth depends on numbers (with the coefficient of density dependent

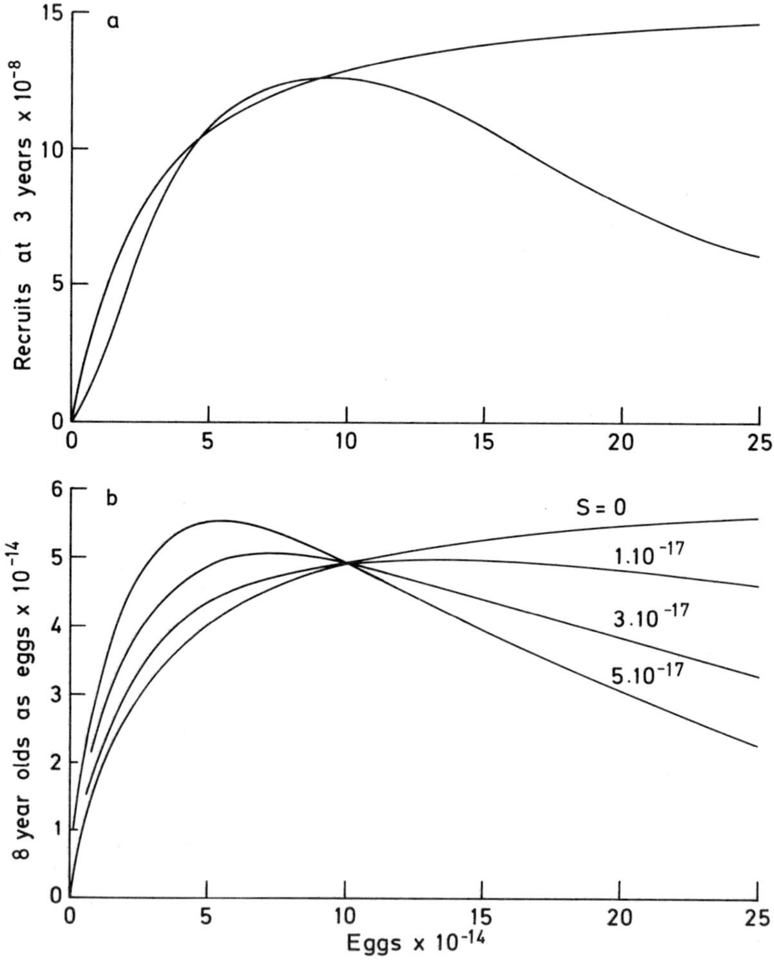

FIG. 4. (a) The asymptotic dependence of recruitment, in numbers, on stock (as eggs) with growth density dependent, but mortality independent of growth; the dome is that in Fig. 1. (b) The dependence of recruitment, as eggs, on stock (as eggs) with growth density dependent, but mortality independent of growth; density dependence of growth is given in the parameter *S*.

growth, S), but the death rate is quite independent of growth. In the second, growth is again density dependent, but mortality varies inversely with weight (using equation (2)), which approximates to the Ricker–Foerster thesis (that density-dependent mortality is a function of the abundance of food). Figures 4(a) and (b) illustrate the first possibility: the dependence of recruit numbers on stock (as eggs) is asymptotic, but there is a dome in the relation of recruits (as eggs) on stock (as eggs). Further density-dependent

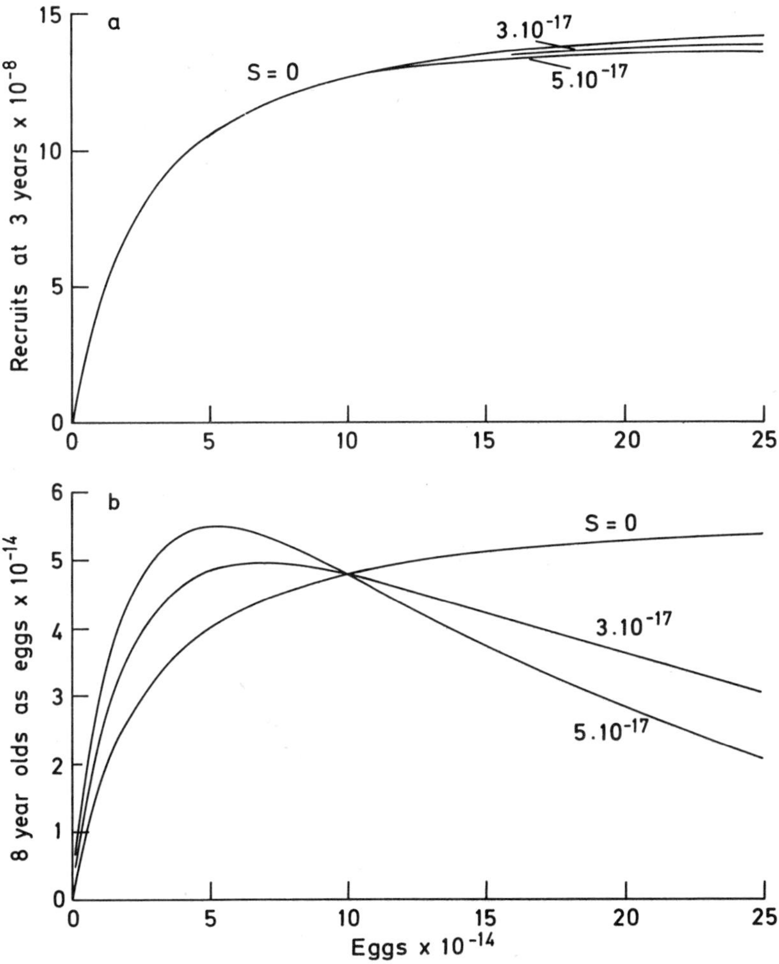

FIG. 5. (a) The dependence of recruitment, in numbers, on stock (as eggs) with growth density dependent and mortality dependent on growth. (b) The dependence of recruitment, as eggs, on stock (as eggs) with growth density dependent and mortality dependent on growth.

growth is observed as a factor of about 0·3 between 0·1 and $1{\cdot}0 \times 10^9$ in stock size. Two of the observational requirements are fulfilled.

Figures 5(a) and (b) illustrate the second possibility, which is effectively the Ricker–Foerster thesis. A weak dome is found in the relation between recruit numbers and the stock of eggs; a pronounced dome is found between recruits (as eggs) and stock (as eggs); and density-dependent growth is observed as a factor of 0·3 to 0·5 between 0·1 and $1{\cdot}0 \times 10^9$ in stock size. All three observational requirements are satisfied.

The model in which growth and mortality are cross-linked satisfies three observations; this is effectively the Ricker–Foerster thesis. That in which growth is density dependent but mortality is independent of growth satisfies two observations. That in which growth and mortality are independent satisfies one observation. The Ricker–Foerster hypothesis is more satisfactory than the other two, but the dependence of recruit numbers on stock in eggs is so weakly domed as not to match the observed curve at all. Further, to produce a dome with the competition parameters, the density dependence of growth becomes much too great, with little effect in fact upon the dome. If growth and density-dependent mortality are ineffective in generating a dome, then some added component of stock-dependent mortality is needed, such as the aggregation of predators or cannibalism.

Figures 6(a) and (b) illustrate a model in which growth is fixed and mortality is stock dependent (as in equation (3)). The dome appears both in numbers of recruits and in recruits (as eggs), which is observed, but there is no density-dependent growth in the system at all. This is the Ricker formulation by which the stock and recruitment relationship is explained entirely by the aggregation of predators at an early stage in the life history. When the fry of the Pacific salmon leave the lakes and move downstream, trout aggregate there and eat the fry; the number of predators varies with the numbers of fry. However, Fig. 2(b) shows that there is density-dependent growth in the young sockeye salmon.

The argument suggests a further possibility—that growth and mortality should be linked in the Ricker–Foerster manner and that there should be an additional component of stock-dependent mortality:

$$\frac{dN}{dt} = -M_1 N - \frac{M_3}{W_t} N^2 - M_4 N_0 N.$$

We have not illustrated this model because the figures would be identical with Figs. 6(a) and (b).

The aggregation of predators may well provide the stock dependence in the Pacific salmon and the Ricker–Foerster mechanism may account for the observed density dependence of growth. However in the sea, fish larvae are somewhat dispersed, at a density of 1–5 m^{-3}. If the rate of aggregation depends on density, the predators must gather slowly, yet if growth and mortality are linked in the Ricker–Foerster manner (which must be so if growth is density dependent), stock dependence might disappear in time before the predators had become effective.

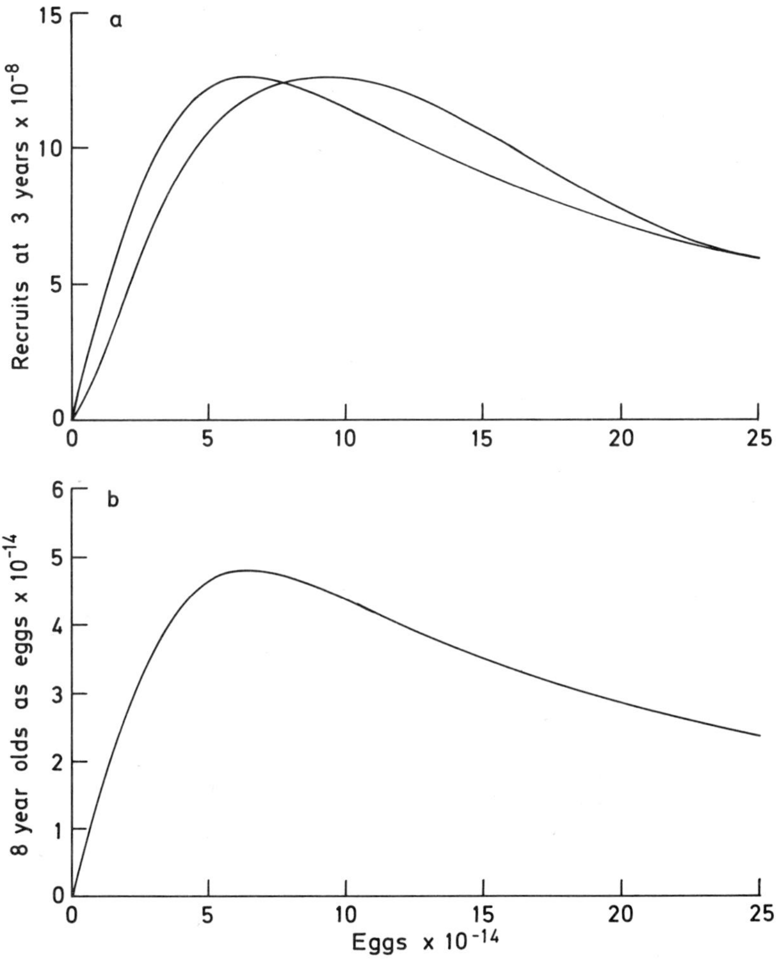

FIG. 6. (a) The dependence of recruitment, in numbers, on stock (as eggs) with growth and mortality independent, but with a component of stock dependent mortality added. (b) The dependence of recruitment, as eggs, on stock (as eggs) with growth and mortality independent, but with a component of stock-dependent mortality added.

It is well known that cod eat young cod and cannibalism has often been invoked as a contributory cause in studies of stock and recruitment. The difficulty has been to distinguish cannibalistic predation from other forms—do the cod aggregate on young cod? or are they merely the constant and indistinguishable predators demanded in the Ricker–Foerster thesis? There are two important points about cannibalism: if stock dependence resides in the numbers of predators only

(a) the process of aggregation is not needed;
(b) the mortality can be inflicted at any age between hatching and recruitment.

There is no need for the aggregation demanded by Ricker and the constant predator in the Ricker–Foerster thesis is already playing its part in the model.

Adult cod eat small cod of 15–20 cm in length, where the feeding grounds and adolescent nursery grounds overlap (Ponamarenko, 1964). Stock in eggs increases from 10×10^{14} to 25×10^{14} eggs (or from 5×10^{8} to $12{\cdot}5 \times 10^{8}$ adults if the average fecundity is 2×10^{6}) as numbers of recruits decline from $12{\cdot}5 \times 10^{8}$ to $6{\cdot}0 \times 10^{8}$. If the increase in death rate is entirely due to cannibalism, it amounts to $\frac{1}{2}$ of a recruit per adult fish, on average. We conclude that cannibalism is a conceivable source of stock-dependent mortality in the gadoids.

4. Discussion

The stock/recruitment curve of gadoid fish is usually dome-like; it is possible that such a dome also occurs in salmon, at least in the Pacific salmon. The analysis here is based on three characteristics observed in the Arcto-Norwegian cod stock:

(a) a dome in the curve of recruits in numbers on stock as eggs;
(b) a dome in the curve of recruits in eggs on stock as eggs;
(c) growth is density dependent, at least amongst immature fish.

The first dome represents the survival of numbers of recruits from the stock of eggs and in the Ricker–Foerster thesis the effect of growth is to moderate that survival. The second dome represents the contribution of new recruitment in biomass added to the stock and the effect of growth is to moderate such additions. The presence of density-dependent growth in the system lends point to the study of the role of growth in the stabilization mechanism in fishes.

Table 1 summarizes the argument. The nature of the link between growth and mortality in the model is shown on the left, and in the body of the Table are given the match to the selected characteristics of the stock and recruitment curve in cod.

TABLE 1. The match of model structures with three observed characteristics of the stock and recruitment curve in cod

Model		Dome in recruit numbers	Dome in recruit as eggs	Presence of density-dependent growth
Growth	Mortality			
Independent	Independent	No	No	No
Density dependent	Not linked to growth	No	Yes	Yes
Density dependent	Linked to growth	Yes	Yes	Yes
Independent	Independent, stock dependence added	Yes	Yes	No

From this Table, the three characteristics are obtained only when growth and mortality are linked to each other, i.e. growth is density dependent and mortality depends on growth. But, in fact, the dome in recruit numbers is only present in a theoretical sense. To generate a pronounced dome a component of stock-dependent mortality is needed, and cannibalism is a better candidate than the aggregation of predators.

We have investigated the link between growth and mortality and the part it plays in the stabilisation mechanism in cod. It would be of considerable interest to extend the analysis to other fishes, for example, plaice and herring, but as the dome is probably absent from the stock/recruitment curves for these animals, the same form of analysis cannot be used. However, there is a remarkable difference between the groups as shown in Table 2. The material on herring was taken from Graham *et al.* (1972) and Bückmann (1942), with a weight/length relation constructed from material in Blaxter and Hempel (1963). The material on plaice was given to me by Mr. Harding and Dr. Lockwood of the Lowestoft Laboratory.

TABLE 2. The daily growth rates of larval cod, plaice and herring

	Percentage weight per day	K	Increment to metamorphosis (70 days)
Cod	10·24	0·0975	×919
Plaice	6·00	0·0583	×59
Herring	4·00	0·0392	×16

After metamorphosis the growth rate of the O-group fish in all three groups is about 2·0–2·5% per day. The difference in growth rates during the larval life is probably enough to account for the marked difference in growth pattern between the species; they are probably genetic differences. The growth of the cod is perhaps excessive, in that part of the gonad may be devoted to cannibalism in order to achieve stabilization.

In plaice and herring, there is no dome in the stock and recruitment relationship and neither fish is cannibalistic in the sense that the adult gadoid is; the adult plaice eats worms and the adult herring is planktivorous. But I-group plaice eat O-group plaice on the beaches and herring larvae find their way into the guts of adult herring. The argument presented here would suggest that as the Ricker–Foerster thesis might be adequate to generate the weakest dome, cannibalism is not needed. But the argument for the cod depends on the observed mortality rates at low stock from the stock/recruitment curve. The same method cannot be extended to solve the problem of whether cannibalism is important in plaice and herring.

References

Beverton, R. J. H. and Holt, S. J. (1957). On the dynamics of exploited fish populations. *Fish. Invest. London*, **19**, 533pp.

Blaxter, J. H. S. and Hempel, G. (1963). The influence of egg size on herring larvae (*Clupea harengus* L.). *J. Cons. Int. Explor. Mer* **28** (2), 211–240.

Bückmann, A. (1942). Die Untersuchungen der biologischen Anstalt über die Okologie der Heringsbrut in der Südlichen Nordsee, *Helgoländer Wiss Meeres* **3**, 1–57.

Cushing, D. H. and Harris, J. G. K. (1973). Stock and recruitment and the problem of density dependence. *Rapp. P.-V. Cons. Int. Explor. Mer* **164**, 142–155.

Dannevig, H. (1889). Aarsberetning om Flodevigens Udklackningsanstalt Virksomhedi Femaaret, 1883–1888.

Foerster, R. E. (1968). The sockeye salmon, *Bull. Fish. Res. Bd Can.* **162**, 422pp.

Garrod, D. J. and Jones, B. W. (1974). Stock and recruitment relationship in the Northeast Arctic cod stock and the implications for management of the stock. *J. Cons. Int. Explor. Mer* **36** (1), 35–41.

Graham, J. J., Chenoweth, S. B. and Davis, C. W. (1972). Abundance, distribution, movements and lengths of larval herring along the Western coast of the Gulf of Maine. *Fish. Bull.* **70** (2), 307–321.

Harding, D., Nichols, J. H. and Tungate, D. S. The spawning of the plaice in the Southern North Sea. ICES Symposium (1975) on the Changes in the North Sea Fish Stocks and their Causes, *Rapp. et P.-V. Réun. Cons. Int. Explor. Mer* (in press).

Harris, J. G. K. (1974). The effect of density dependent mortality on the shape of the stock and recruitment curve. *J. Cons. Int. Explor. Mer* **36** (2), 144–149.

Herrington, W. C. (1948). Limiting factors for fish populations. Some theories and an example. *Bull. Bingh. Oceanogr. Coll.* **11** (4), 229–283.

ICES (1967). Report of Northeast Arctic Working Group. (mimeo).

ILES, T. D. (1968). Growth studies on North Sea herring II. O-group growth of East Anglian herring. *J. Cons. Int. Explor. Mer* **32** (1), 98–116.

JONES, R. (1973). Density dependent regulation of the numbers of cod and haddock. *Rapp. P.-V. Cons. Int. Explor. Mer* **164**, 156–173.

LETT, P. F., KOHLER, A. C. and FITZGERALD, D. N. (1975). Role of stock biomass and temperature in recruitment of Southern Gulf of St. Lawrence cod *Gadus morhua. J. Fish. Res. Bd Can.* **32** (9), 1613–1627.

PALOHEIMO, J. E. and KOHLER, A. C. (1967). Analysis of the Southern Gulf of St. Lawrence cod population. *J. Fish. Res. Bd Can.* **25** (3), 555–578.

PONOMARENKO, I. Y. (1964). Effects of the hydrological regime and feeding conditions on the formation of the abundance of a yearclass of cod during the life of its yearlings in the near bottom layers of the Barents Sea. (Trudy *PINRO*), **16,** 235–249.

RAITT, D. F. S. (1968). The population dynamics of the Norway pout in the North Sea. *Mar. Res.* (5), 24pp.

RICKER, W. E. (1958). Handbook of computations for biological statistics of fish populations. *Bull. Fish. Res. Bd Can.*, **119**, 300pp.

RICKER, W. E. and FOERSTER, R. E. (1948). Computation of fish production. *Bull. Bingh. Oceanogr. Coll.* **11** (4), 173–211.

SAETERSDAL, G. and CADIMA, E. (1960). A note on the growth of the Arctic cod and haddock. ICES Gadoid Fish Committee, Paper 90, 3pp. (mimeo).

SAHRHAGE, D. and WAGNER, G. On fluctuations in the haddock population of the North Sea. ICES Symposium (1975) on the Changes in the North Sea Fish Stocks and their Causes, *Rapp. et P.-V. Réun. Cons. Int. Explor. Mer* (in press).

SOUTHWARD, G. M. (1967). Growth of Pacific halibut. *Int. Halibut Commn* **43**, 40.

3. Growth, Reproduction and Mortality in Gadoid Fish Species

R. Jones and C. Johnston

Marine Laboratory, Aberdeen

1. Introduction

This paper describes attempts that have been made to model the growth of fish. It is assumed that growth is able to take place when the rate of energy intake is more than sufficient to supply the energy needed for maintaining an organism at its existing size and according to its normal mode of existence.

Investigations starting from this viewpoint were found to lead to two lines of enquiry.

(1) What is the relationship between rate of food intake and body weight necessary to account for growth and reproduction in particular species? Hence, by what biological process has this been achieved?

(2) Given that a fish takes in more energy than it needs for maintenance what would be the effect of deploying this energy in different ways? There would appear to be a choice, in an evolutionary sense, of whether to use surplus energy for growth, reproduction, laying down storage tissue, or for more energetic food gathering. Natural selection has presumably favoured the genotype that is adapted to deploying energy into these various categories in the optimum way throughout life.

By "optimum way" could be meant the method of using energy that, in the long term, has tended to lead to the maximum number of surviving offspring. As a first approximation, this could be thought of as the method of using energy that leads to the maximum diversion of energy into reproductive products throughout the entire course of the life of a cohort of individuals. One is prompted to ask questions such as "Why is it an advantage for an Arctic cod to be large and mature late in life?" On the other hand, "Why is it also an advantage for a Norway pout to be small and to mature early?" Presumably both species are successfully adapted to their environments. Given what one knows about these species what can one deduce about the differences between the environments or "niches" in which they live?

Section 2 gives details on which computer simulations have been based. Section 3 deals with the effect on the growth curve of varying the relationship between the rate of food intake and body size in this model. Section 4 deals with the effect on growth of varying the amount of energy going into reproduction. In Section 5, an attempt has been made to deducc something about the natural mortality rate in particular stocks of haddock and cod.

2. Computer Simulations

As a basis for modelling it has been assumed that the energy available for growth and reproduction is simply the difference between the energy intake in the form of food and the energy expended on maintenance. The rate of conversion of energy into growth and reproduction will then depend on this difference and on a factor for conversion efficiency, i.e.

$$GR + RE = [FE - ME]\mathscr{E}$$

so that

$$GR = [FE - ME]\mathscr{E} - RE, \tag{1}$$

where

GR = growth rate,
FE = rate of energy intake (food energy),
ME = rate of energy consumption for normal "active" maintenance (maintenance energy),
RE = rate of increase in the energy content of the maturing gonad (reproduction energy)

(these terms are all measured in energy units (i.e. kcalories)), and

$\mathscr{E}$ = the efficiency of conversion of available energy into growth or gonad energy.

Each of these terms is of course a function of body size, and appropriate expressions are described below. Using these, a computer simulation has been written to model growth and reproduction. Growth, for individual fish, has been simulated for a lifetime of 20 years, each year consisting of 100 steps. Reproductive output, for cohorts of fish, has been simulated for the entire life span, allowing for the effects of different rates of mortality. Further details are given in the Appendix.

Growth energy

Energy required for growth can be determined directly from the growth rate of fish at a given size. Figures 1 and 2 show the annual growth increments of cod and haddock plotted as a function of body weight, and indicate that the growth rates of these species are extremely variable throughout the North Atlantic. It is quite clear from these results that there is no such thing as "a cod growth rate" or "a haddock growth rate". Thus, it is necessary to simulate the growth process in particular stocks of cod or particular stocks of haddock.

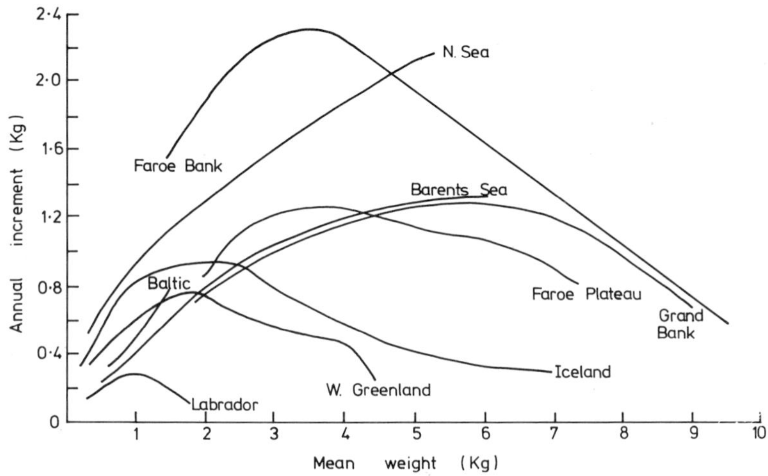

FIG. 1. Relationship between annual growth increments (kg) and mean weights (kg) for north Atlantic cod.

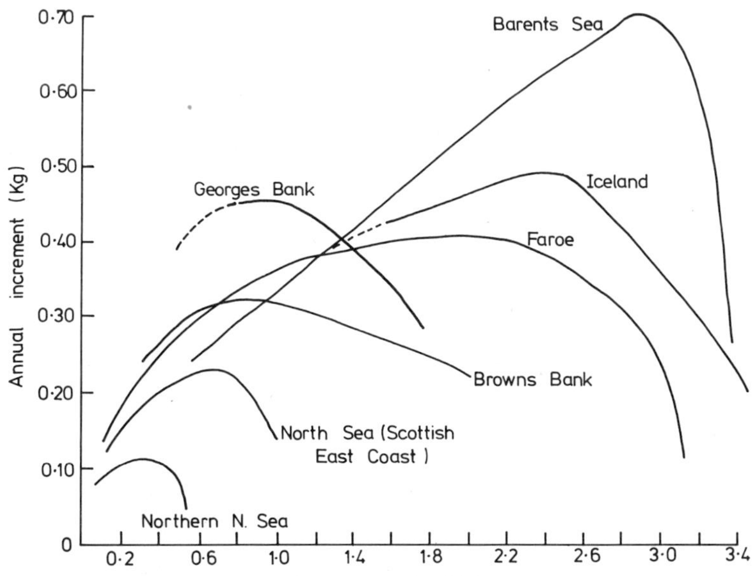

FIG. 2. Relationship between annual growth increments (kg) and mean weights (kg) for north Atlantic haddock.

Maintenance energy

For a very large number of organisms, maintenance energy has been found to be proportional to W^n where n is about 0·7–0·8 (Zeuthen, 1953; Winsberg, 1956). In this study an exponent of 0·8 has been used and the factor of proportionality can be written as a function of temperature and swimming speed, i.e.

$$\text{ME} = 0{\cdot}007\,W^{0{\cdot}8}\exp(0{\cdot}081T+0{\cdot}76V) \text{ kcalories per day at } T\,^{\circ}\text{C}.$$

To allow for the energy required for maintaining the developing gonad, an additional term was added to this equation to give

$$\text{ME} = 0{\cdot}007\,W^{0{\cdot}8}\exp(0{\cdot}081T+0{\cdot}76V) + 0{\cdot}007\,Wg^{0{\cdot}8}\exp(0{\cdot}081T) \text{ kcalories per day,}$$

where

W = body weight excluding gonads,

Wg = weight of gonad,

T = temperature in °C,

V = swimming speed in body lengths s^{-1}.

It is assumed in this equation that maintenance is the same for the gonad as for the rest of the body.

Reproduction energy

Energy for reproduction can be calculated in various ways. For females, for example, one could calculate the total energy content of the mature gonad or of the eggs released. For some gadoid fish species the mature gonad is approximately 10% of the body weight. If one assumes that the calorific value of a gadoid ovary is the same as that of a roach ovary, one can use a value of 1·7 kcalories g^{-1} from Mann (1965), to determine the energy content of a mature gonad. If the calorific value of normal body mass is assumed to be 1·0 kcalories g^{-1}, this leads to a value of 17% as the energy value of a mature gadoid ovary expressed as a percentage of the somatic energy in the body at the time of spawning.

An alternative approach is to determine the calorific value of the total eggs produced during the course of the spawning season. Table 1 shows some estimates based on this method of calculation. For example, a 5 kg cod produces about 2 million eggs. Using data from Daan (1975), the energy content of these eggs would be about 866 kcalories and this would represent about 22% of the somatic energy of the body at the time of spawning.

Similar calculations for herring using data from Blaxter and Hempel (1963) give values of 19–32% for the energy content of the eggs.

TABLE 1. Showing the energy content of the eggs produced by three species

Species	Body wt (g)	Fecundity	Dry wt of 1 egg (mg)	Dry wt of all eggs (g)	Energy content of eggs (kcalories)	Energy content of body	Energy content of eggs as % of somatic weight
Herring A[1]	110	22 900[2]	0·242[3]	5·55	28[4]	150[5]	19
B	180	49 000[2]	0·317[3]	15·5	78[4]	245[5]	32
Cod	5000	2 038 000[9]	0·0746[6]	152	866[7]	4000[8]	22
Whiting (aquarium spawning, 1972)	592[10]	1 140 000[10]	0·037[12]	42	211[4, 10]	533[8, 10]	40

Notes

(1) A: 3-year-old Clyde herring in 1961; B: 4-year-old Clyde herring in 1962. (Baxter, personal communication.)
(2) Calculated using fecundity/weight relationship of $F = 373W - 18\,100$ from Baxter and Pope (1969).
(3) From Blaxter and Hempel (1963).
(4) Calculated assuming 5 kcalories g^{-1} dry weight of eggs.
(5) Calculated assuming 1·36 kcalories g^{-1} wet weight of body (Daan, 1975).
(6) From Oosthuizen and Daan (1974).
(7) Calculated assuming 5·7 kcalories g^{-1} dry weight of eggs (Daan, 1975).
(8) Calculated assuming 0·9 kcalories g^{-1} wet weight of body (Daan, 1975).
(9) Oosthuizen and Daan (1974) give a fecundity/weight relationship of $F = 569W - 807\,000$.
(10) Data for two female fish (31 and 32 cm) combined from Hislop (1975).
(11) Calculated assuming 0·816 kcalorie g^{-1} wet weight (Daan, 1975).
(12) From Hislop (1975).

It is clear that, quite apart from the energy required for such activities as spawning behaviour and spawning migrations, the amount of energy required for egg production must be relatively quite large. For purposes of calculation it has been assumed that the mature ovary has an energy content of 20–30% of the somatic energy at the time of spawning.

It has been assumed that the weight of the ovary at the start of gonad production depends on the body weight but that it then grows independently. If W^1 is the weight of a mature fish at the time when it starts producing gonad, then at time t later the weight of gonad is given by

$$Wg = \alpha_3(1+p)^t W^1.$$

Using cod data from West (1970), α_3 has been taken as 0·008, while t has been put equal to 50 (this is equivalent to 6 months). For each simulation a value of p has been chosen to make the final energy content of the mature gonad some required percentage of the somatic energy at the time of spawning.

Conversion efficiency

To determine the food energy required for observed growth rates, a conversion efficiency has been calculated for gadoids. It appears that the conversion efficiency for growth declines slowly with body weight, the relationship being

$$\text{conversion efficiency } \mathscr{E} = 0{\cdot}79\,W^{-0{\cdot}15}, \quad 5\text{ g} \leqslant W \leqslant 900\text{ g}. \tag{2}$$

This relationship has been determined using data for fish in the range 5–900 g, but it has also been used in the simulation for heavier fish. Estimates of the efficiency of conversion of energy into reproductive products have been made using experimental data supplied by Hislop, given in Table 2. This gives details of the body weights, growth, food intake and egg production of three whiting that matured their ovaries and spawned in an aquarium during autumn and spring 1969/70.

From these data, the energy available for reproduction has been calculated from the relationship food energy − maintenance energy − energy needed for growth. In terms of equation (1) this is equivalent to writing

$$\text{RE}/\mathscr{E}_1 = \text{FE} - \text{ME} - \text{GR}/\mathscr{E},$$

where $\mathscr{E}_1$ is the efficiency of conversion of energy into reproductive material and $\mathscr{E}$ is the efficiency of conversion of energy into growth. An estimate of $\mathscr{E}_1$ was then obtained from the ratio of energy content of liberated eggs/(RE/$\mathscr{E}_1$).

TABLE 2. Estimation of the efficiency of conversion of energy into eggs (data for three fish in 1970, from Hislop, 1975)

Mean body weight (W)	Growth (G)	Total food (g)	Energy (kcalories) Food A	Maintenance[(1)] B	Growth[(2)] C	Reproduction[(3)]	Eggs	Dry weight of eggs (g)[(4)]	Energy content of eggs[(5)]	Conversion efficiency[(6)]
987	−267	1052		445	−954					
594	−34	780		296	−112					
363	−29	244		200	−88					
1921		2076	2076	941	−1154	2289	2334	93·4	532	0·23

Notes

(1) Maintenance energy at 10 °C for 94 days = $1{\cdot}79 W^{0{\cdot}8}$ kcalories (calculated for each fish separately).
(2) Energy for growth = $G/0{\cdot}79 W^{-0{\cdot}15}$ kcalories (calculated for each fish separately).
(3) Energy for reproduction = $A - B - C$.
(4) Using conversion factor of 0·04 mg per egg dry weight.
(5) Using conversion factor of 5·7 kcalories g^{-1} dry weight of eggs.
(6) Energy content of eggs/energy available for reproduction.

The value obtained for conversion of energy into eggs was 0·23. For comparison, the growth efficiencies for the fish used in the experiment, calculated using equation (2), ranged from 0·28 to 0·33.

In view of the approximate nature of the calculation in Table 2, it has been concluded that the egg conversion efficiency was not significantly different from the growth efficiency.

For the simulation, therefore, equation (2) has been used to determine the efficiency of conversion of energy into both growth and gonad energy.

Food energy

Provided sufficient information is available about growth rates, maintenance energy requirements and the energy needed for reproduction, it is possible to determine the food energy requirements throughout the life of a fish which accounts for an observed relationship between size and age. Preliminary estimates of energy consumption as a function of body weight can be determined for North Sea and Faroe cod and for North Sea and Faroe haddock. In each instance it has been found that the rate of total energy intake increased with body weight and that this could be approximated by taking the rate of energy intake proportional to W^r where W is the weight of the body and r is a constant.

Although a perfect logarithmic relationship did not hold over the entire life span it was decided to use this simplification as a first approximation. This left a choice of whether to treat the rate of energy intake as a function of body weight or body length. Immediately after spawning, for example, somatic weight is normally lower than it was 6 months previously, whereas body length is not. The choice of body weight or body length (L) for calculating the rate of energy intake therefore led to different results, although in practice it was not found to affect any of the conclusions. In the calculations described here it has been assumed that

$$\mathrm{FE} = \lambda \alpha_1 L^{r_1},$$

where λ is a parameter introduced to take account of differences in the summer and winter rates of feeding (see Appendix).

3. The Relationship between Food Intake and the Shape of the Growth Curve

This relationship has been investigated by varying the parameters α_1 and r_1 in the part of the model dealing with rate of energy intake, i.e.

$$\mathrm{FE} = \alpha_1 \lambda L^{r_1}.$$

In the case of the parameter α_1, variations were found to affect the maximum body size attained (Fig. 3). In general the larger the value of α_1

the larger is the body size attained after any fixed time period. The lower growth curve resulted from a reduction in α_1, and hence in food intake, of 12%, indicating how sensitive growth could be to fluctuations in rate of food intake. The cyclical fluctuations in the curves are a result of the cyclical effects of reproduction on growth and body weight, associated with annual variations in condition factor.

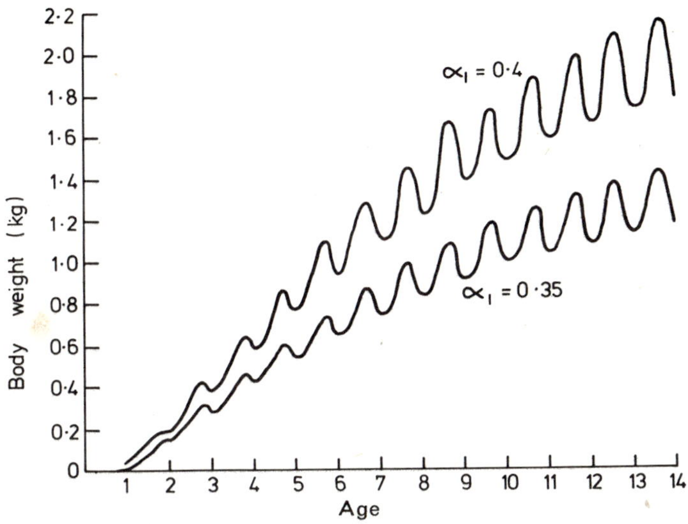

FIG. 3. Effect on growth of variations in the coefficient α_1.

Figure 4 shows examples of ways in which variations in r_1 affect the "shape" of the growth curve (shown here as body length plotted against age). In general, the larger the value of r_1 the longer the time required to approach the asymptote. The seasonal fluctuations in length are not as marked as those for weight (Fig. 3) because it is assumed that spawning leads to a cyclical loss in weight, but not in length. For comparative purposes, the parameters in the Bertalanffy growth equation

$$L = L_\infty\{1 - \exp[-K(t - t_0)]\}$$

have been fitted to the growth curves generated by the simulation model. The results can be summarized as follows.

(1) Variations in α_1 are positively correlated with variations in L_∞. They were found to have comparatively little affect on the Bertalanffy parameter K.

(2) Variations in r_1 were negatively correlated with the Bertalanffy parameter K. For example with a value of $r_1 = 1{\cdot}95$, the corresponding

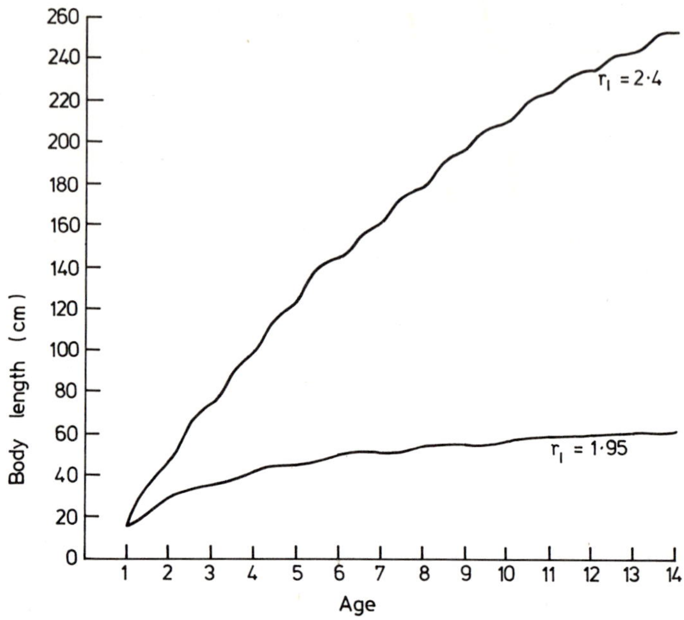

FIG. 4. Effect on growth of variations in the coefficient r_1.

value of K was 0·26. On the other hand, with a value of $r_1 = 1{\cdot}65$ the corresponding value of K was 0·4.

The values of α_1 and r_1 required to generate the growth curves of certain fish species are shown in Table 3. (These include values for Norway pout,

TABLE 3. Showing approximate values of key input parameters for various fish species

Species	Parameters for model			Corresponding Bertalanffy parameters	
	α_1	r_1	Age of first maturity	K	L_∞
Hypothetical species	0·35	0	1	3·2	9·4
Norway pout	0·28	1·65	2	0·4	23
North Sea haddock	0·38	1·95	2–3	0·26	58
Arcto-Norwegian cod	0·13	2·55	8	—(1)	—(1)

Note
(1) The growth curve of Arcto-Norwegian cod is not of the "Bertalanffy form."

North Sea haddock and Arcto-Norwegian cod.) The corresponding relationships between rate of food intake and body size are shown in Fig. 5. It should be noted that since linear relationships on the log/log scale have been assumed, misleading results might be obtained if any of the lines are extrapolated.

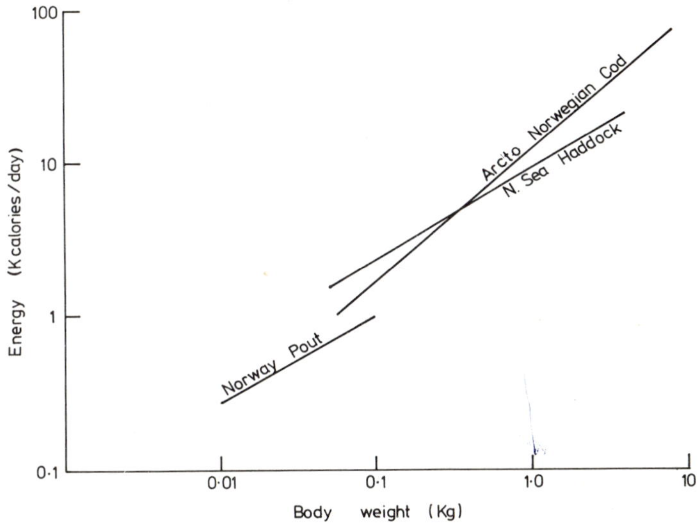

FIG. 5. Showing the daily energy requirements of certain species.

Also shown in Table 3 is one result of simulating growth when $r_1 = 0$ corresponding to the situation in which the rate of food intake is constant with body size. In this instance the value of *K* for the best fitting Bertalanffy curve was found to be 3·2. It is interesting to note that one of the largest values of *K* recorded by Beverton and Holt (1959), in their review of Bertalanffy growth parameters for a large number of fish species, was 3·7 for *Labidesthes*. The possible significance of this will be discussed below.

Some biological implications

Mathematically, the simulation shows that the value of the Bertalanffy parameter *K* is dependent on the way in which the rate of food intake changes with body weight. It is appropriate to enquire, therefore, how different species are able to increase their energy intake with body weight at different rates.

Evidence on this point comes from stomach content analyses, which show that both food particle size and also the nature of the diet may change with body weight. Cod and haddock, for example, have been shown to eat larger food organisms as they increase in size (Jones, in press; Daan, 1973). Cod with body weights less than 10 g consume small crustaceans such as copepods or amphipods. The diet changes to larger crustaceans, to small fish and eventually to large fish. For haddock, the diet also starts with small crustaceans, followed by larger crustaceans. When the body weight reaches about 50 g haddock begin to take up the demersal habit and to exploit benthic organisms. In some parts of their range the largest haddock exploit

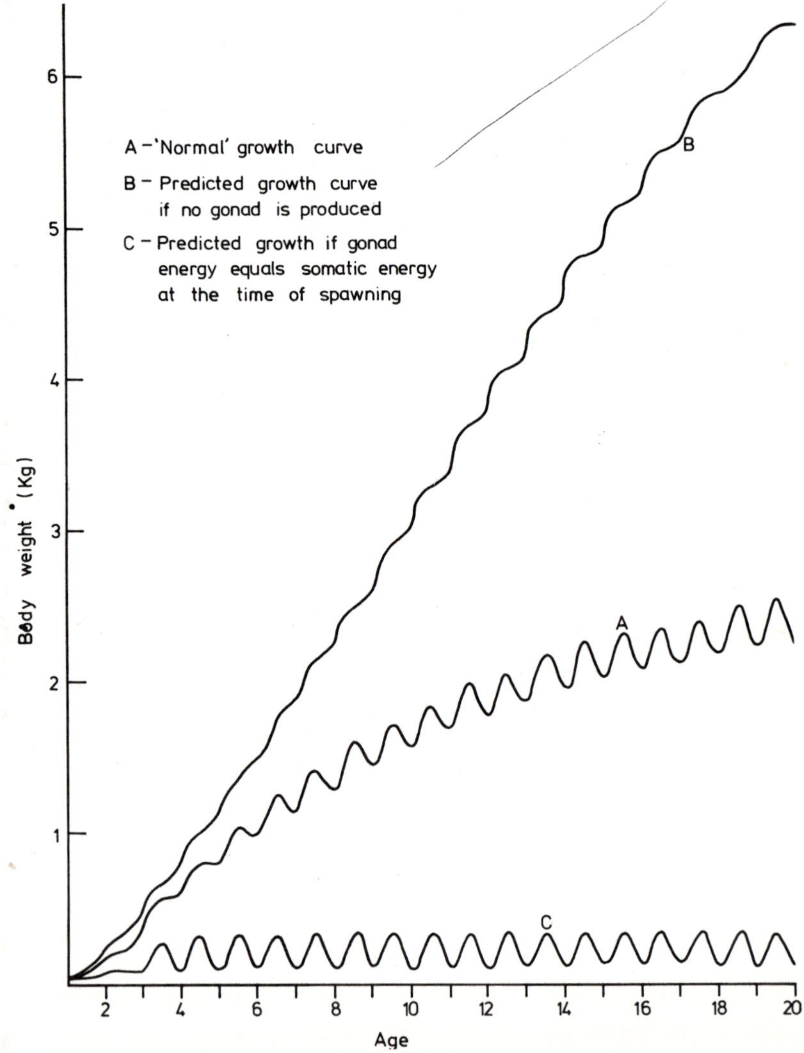

FIG. 6. Effect on growth of varying the growth of the gonad.

fish. Norway pout tend to eat small crustaceans and when fully grown to exploit larger crustaceans.

The implication appears to be that each species occupies, not a single food niche, but a series of food niches. The maximum size attained by a species is presumably determined by the optimum size for exploiting the final food niche. The Arcto-Norwegian cod grows to an appropriate size for exploiting fish such as capelin. Possibly individuals continue to grow and exploit a series of food niches until a discontinuity is reached. An excellent example is given by Deelder (1951) with reference to perch. He mentions that at a length of about 13 cm, perch change from feeding on small invertebrates to feeding on fish. In lakes where there is sufficient weed to afford protection for such small fish, the perch grow normally at first, but stop growing at about 14 cm. In lakes, however, where small fish are more accessible the perch grow to a much larger size.

In the case of species such as *Labidesthes* one might postulate that either: (a) there is only one food niche, exploited with equal efficiency by individuals of all sizes; or (b) there is a succession of food sources, and that it so happens that the rate of energy intake is constant with increasing body weight. The conclusion from this section is that for some species, food energy is derived from a succession of food sources. The final shape of the growth curve is then determined partly by the rate at which food intake changes with size and partly by the optimum size for exploiting the final food niche.

4. The Relationship between Reproduction and the Shape of the Growth Curve

Section 3 of this paper has been concerned with the relationship between food intake and body weight in certain species. This section is concerned with possible variations in the growth of these species if they utilized their food energy in different ways; in particular, what might happen to the growth of a cod or a haddock if it varied the amount of energy going into reproduction. This could occur if they changed either the size of the gonad or the age of first maturity.

The effect of varying gonad size

Given a relationship between food intake and body weight for a species such as North Sea haddock it is a simple matter to simulate growth curves for fish producing gonads of different size (Fig. 6). A "normal" haddock (curve A) attains a weight of 2–3 kg and produces a gonad with an energy content about 20–30% of the energy content of the somatic body at the time of spawning. Curve B shows that if this fish did not spawn at all, but simply converted all its surplus energy into growth, it ought to grow to a very much larger size. At the other extreme, curve C shows the growth of a fish which produces a gonad with as much energy as the energy in the somatic part of the body at the time of spawning. Such a fish ought not to grow larger than 250 g and its condition factor should decline annually by 50% at each spawning.

The effect of varying the age at first maturity

Apart from varying the size of the gonad, an alternative adaptation would be to vary the age of first maturity. A "normal" haddock (Fig. 7, curve A) is assumed to spawn first at an age of 2 years. Curves B and C show the effect of postponing the age of first maturity to 7 and 13 years, respectively. All three curves converge to the same point at infinite age. It is interesting to note that if first maturity is postponed sufficiently it would be theoretically possible for a fish to grow so large that reproduction when it occurred would

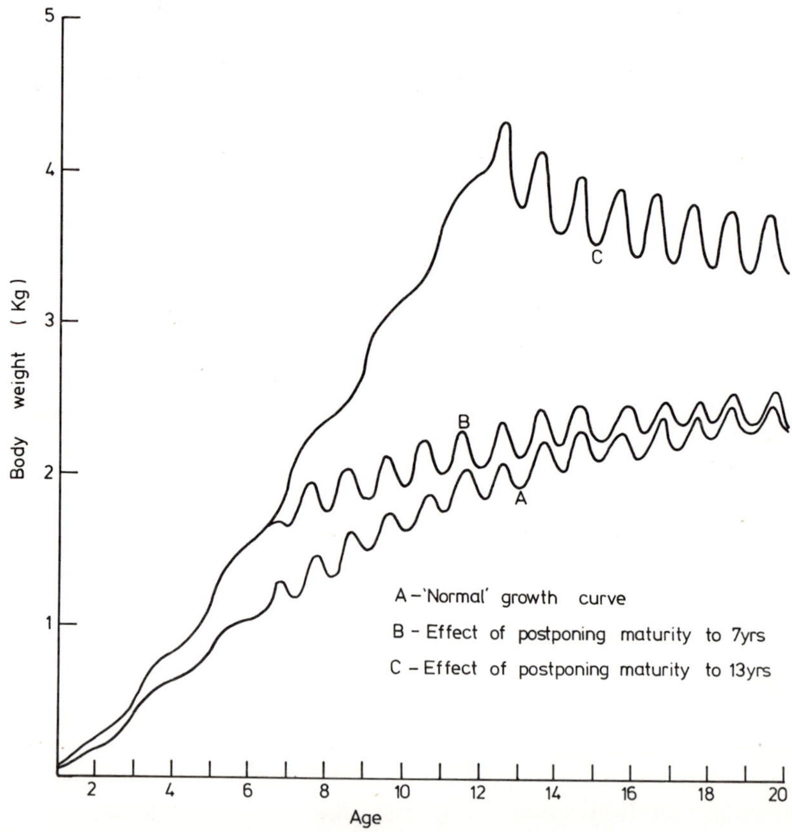

FIG. 7. Effect on growth of varying age at first maturity.

lead to a gradual decline in body weight and hence in condition throughout the remainder of the life.

These results show that because of the high energy cost of egg production, reproductive strategy could have a significant effect on growth. Figures 6 and 7 indicate just how large the effects might be.

Relationship between gonad size and food availability

One of the applications of a growth model is in simulation studies of predator–prey interactions. In this situation a difficulty is to know what kind of predator response to use, when there is a shortage of prey.

There are various possibilities:

(1) Fish might be adapted to survive at the expense of growing more slowly. Figure 3 shows how potentially sensitive growth could be to variations in the rate of food intake. Although haddock and cod show variations in growth rate, these do not appear to have been as large as might have been expected on this hypothesis.

(2) Fish might be adapted to keep growing and producing eggs in a predeterminal way, even if they died in the process. For haddock and cod evidence on this possibility is inconclusive. If fish were adapted to grow and to produce gonads of a predetermined size, then mature fish would tend, in times of food scarcity, to have lower condition factors than normal. If this were the case, one might have expected to find smaller condition factors in mature individuals taken from a large year-class than from a small one. There appears to be little, if any, evidence to support this possibility.

(3) Mature fish might be adapted to conserve energy by producing a smaller gonad than average. Thus they might be adapted to prevent too great a loss in somatic energy by drawing on the gonad energy as a reserve when food is scarce. Figure 6 shows how potentially sensitive growth might be to variations in gonad size.

There is evidence from the literature to support this possibility. Tyler and Dunn (1975) investigated the relationship between food intake and ovary condition in winter flounder. They found that fish kept on small rations showed a smaller proportion of yolked oocytes and concluded that this arose from a reduction in the number of oocytes starting vitellogenesis. Similar investigations by Scott (1962) on rainbow trout showed that on low rations fewer eggs were brought to maturity. He considered that the greater the starvation, the higher the proportion of maturing follicles which regressed. Macer (1974) working on horse mackerel stated that the oocytes could be resorbed both before and after spawning. Bennett and Gibbons (1975), working on largemouth bass, considered that resorption of unshed eggs could occur at the end of the spawning season. De Veen (in press), investigating sole, concluded that partially developed gonads could be resorbed at the end of the spawning season.

Annual differences in fecundity or ovary size for fish of a given age and length have been reported by several workers (Hodder (1965) for Grand Bank haddock, Richards (private communication) for Norway pout and Martin (1970) for lake trout). Fry (1949) observed that even after maturity, it is not unusual in some seasons for some lake trout to fail to produce gametes.

There would appear to be considerable evidence therefore to support the theoretical possibility that when food is scarce, energy is conserved at the expense of the developing gonad.

5. Natural Mortality and Reproduction

Although the process of natural mortality is not properly understood it has been suggested that individuals may be more vulnerable to predation or disease at, or just after, spawning (Orton, 1929). To investigate this relationship estimates have been made of the total energy going into reproduction throughout the life of a cohort of individuals for different rates of mortality and for different reproductive strategies.

The relationship between gonad size and mortality rate

Consider the gonad production for two genotypes represented by curves A and C in Fig. 8. Curve A is the condition for the "normal" North Sea haddock-type fish in which gonad energy is 20–30% of somatic energy at the time of spawning. Curve C shows the situation for the fish in which the

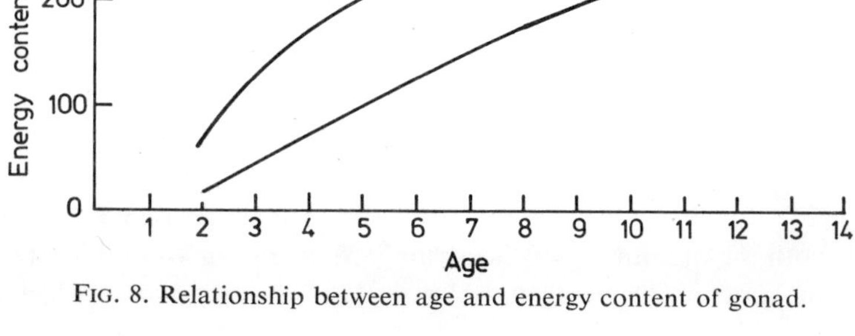

FIG. 8. Relationship between age and energy content of gonad.

energy of the gonad is equivalent to the somatic energy of the fish at the time of spawning. Clearly with a very high mortality rate and a relatively short life it would be better to have a large gonad and produce as many eggs as possible at each spawning. On the other hand, with a low enough

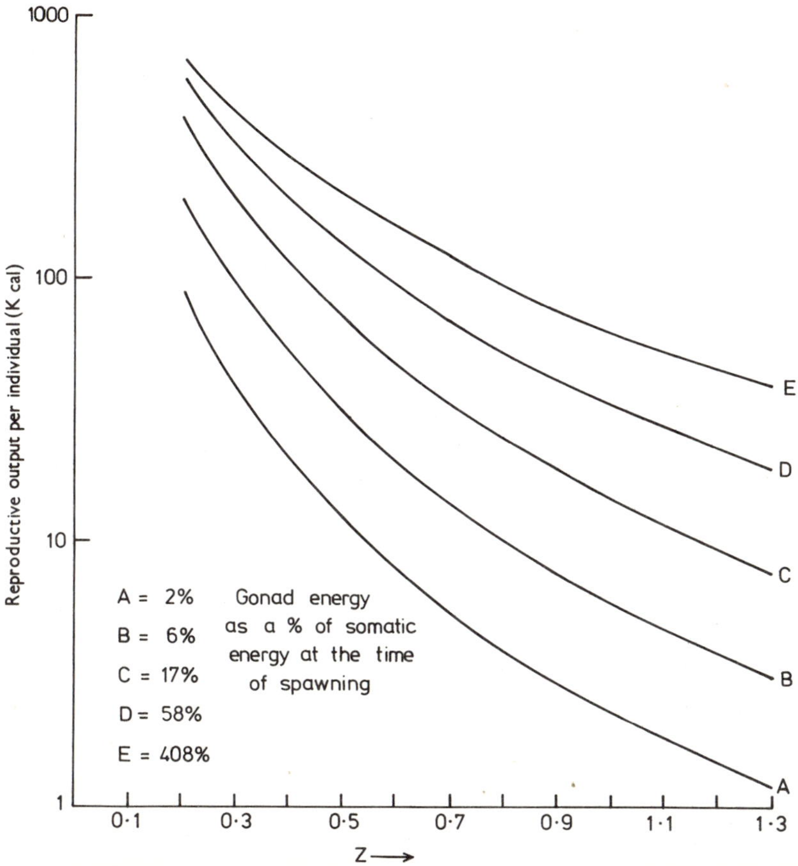

FIG. 9. Relationship between the energy content of reproductive products per individual and the mortality rate for various sizes of gonad expressed as percentage of somatic energy at the time of spawning: A = 2%; B = 6%; C = 17%; D = 58%; F = 408%.

mortality rate there might be an advantage in a proportionately smaller gonad to benefit from the larger body size that might be attained. In each instance the total reproductive output per individual or per group of individuals will depend on the mortality rate adopted. The results of a large number of simulations (Fig. 9) show some relationships between mortality rate (Z) and the expected reproductive output for an individual fish. The curves are for fish with different gonad sizes and they decline because an increase in mortality rate decreases the expected reproductive output per individual whatever the gonad size. Also shown is that for any one mortality rate ($Z \geqslant 0{\cdot}2$) total reproductive output increases theoretically with gonad size up to gonad sizes very much larger than those ever observed.

If it is assumed that present-day genotypes are the most successful of all theoretically possible genotypes it is necessary to explain why fish such as cod and haddock do not produce much larger gonads than they do. A likely explanation is that the larger the gonad, the greater the decrease in condition

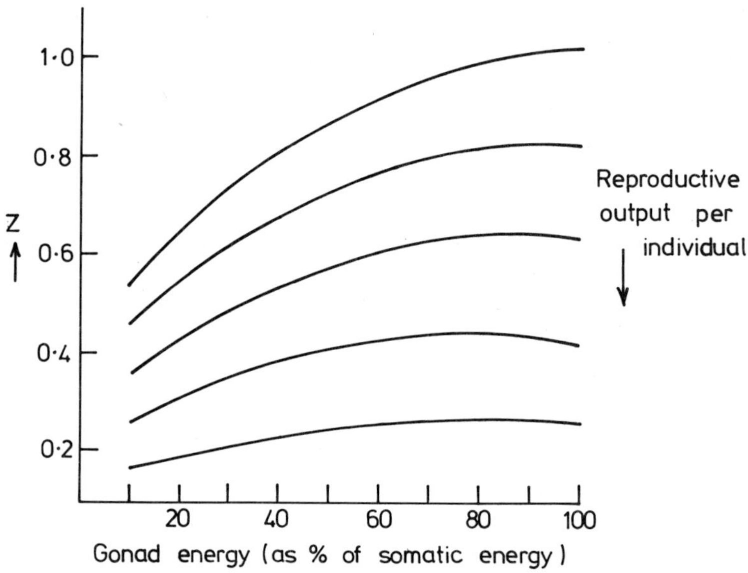

FIG. 10. Relationships between the size of the gonad and the mortality rate for different amounts of energy going into reproduction (North Sea haddock-type growth with first maturity at age 2 years).

factor, and hence the larger the natural mortality rate. A fish growing according to curve A in Figs. 6 and 8 would lose about 20% of its somatic energy at each spawning. The corresponding loss for curve C would be 50%. From the relationships in Fig. 9 it is easy to calculate how much the mortality rate would have to be increased to offset the apparent advantage of producing larger gonads. Figure 10 shows the relationship between gonad size and mortality rate that is necessary to make total reproductive output just independent of gonad size. For example, to a first approximation the apparent advantage of increasing gonad energy from 30 to 40% of somatic energy would be lost if mortality rate increased by only 10%.

Depending on the size of gonad produced, there is a drop in somatic energy at each spawning. The same data used for Fig. 10 have been replotted to show (Fig. 11) the relationship between drop in somatic energy and mortality rate. These two figures summarize the calculations that have been made to relate gonad size to mortality rate. The investigation does not lead to an estimate of the mortality rate for any species, but simply determines the relative magnitude of the effect.

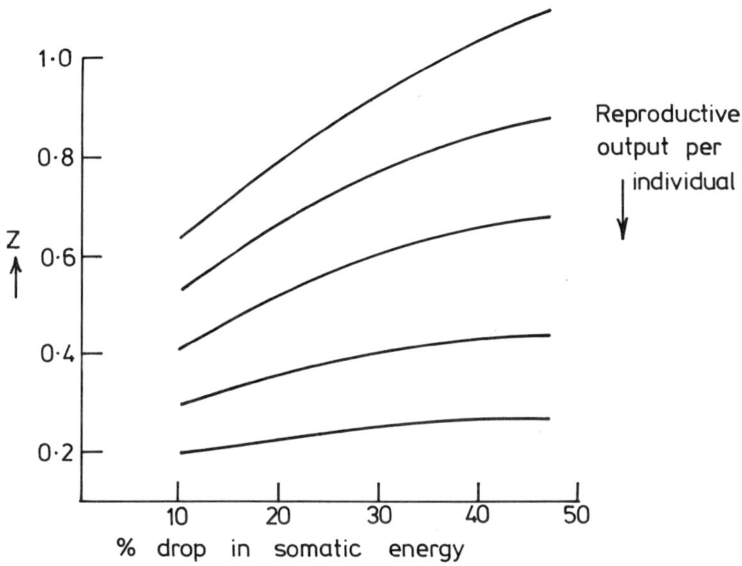

FIG. 11. Relationship between seasonal drop in somatic energy and the mortality rate for different amounts of energy going into reproduction (North Sea haddock-type growth with first maturity at age 2 years).

The relationship between age (or size) of first maturity and mortality rate

Figure 12 summarizes estimates of the total reproductive output of North Sea haddock for different mortality rates and ages of first maturity. If the total mortality rate is as low as 0·05, reproductive output is maximized by postponing maturity to at least 10 years. At the other extreme, if the instantaneous rate of mortality is 0·3, maximum reproductive energy should be attained by maturing at 2 years.

If it is supposed that in the virgin stock the age of first maturity was 2–3 years, as it is today, it can be argued that the natural mortality rate should have been not less than 0·2 or 0·3, otherwise the age of first maturity ought to have been greater than 2–3 years.

The results of similar calculation for Arcto-Norwegian cod (Fig. 13) show the total energy going into reproduction in relation to the age of first maturity for different mortality rates. For example with a value of Z equal to 0·1, reproductive output should be maximized by maturing at about 16 years. On the other hand, if Z equals 0·3 the optimum age of first maturity

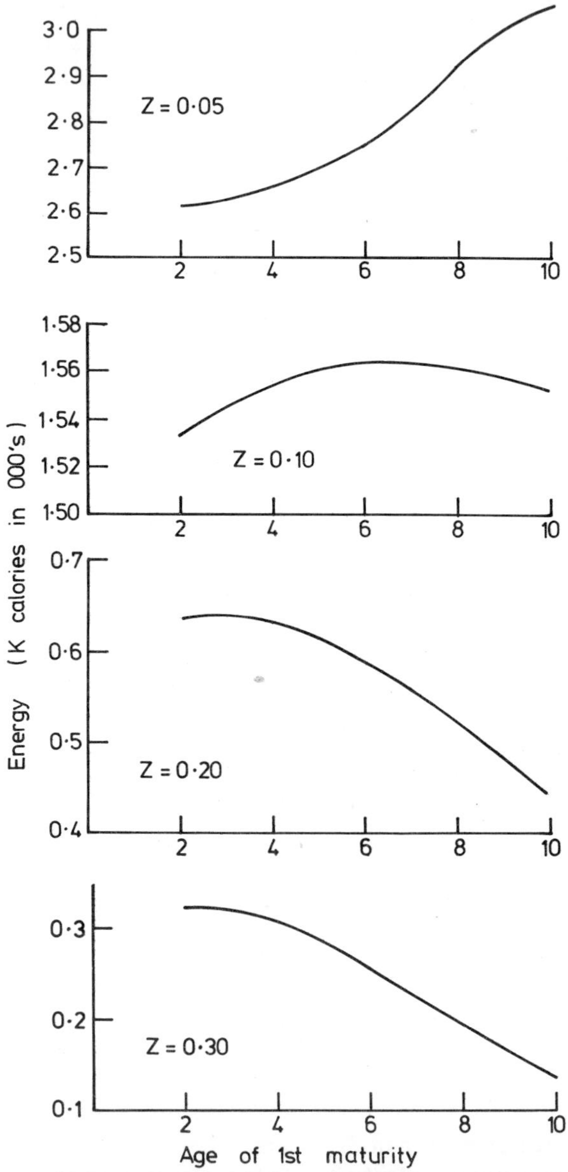

FIG. 12. Relationship between reproductive output per individual and age at first maturity for various values of Z (North Sea haddock-type growth).

should be about 8 years. For this species, the 50% age of first maturity happens to be about 8 years at present, although in earlier years, when the stock was less heavily exploited, the age of first maturity was considerably in excess of this (about 12 years). The results suggest that the natural mortality rate is unlikely to have been greater than 0·2–0·3, otherwise the age of first maturity in the unexploited stock ought to have been less than 12 years.

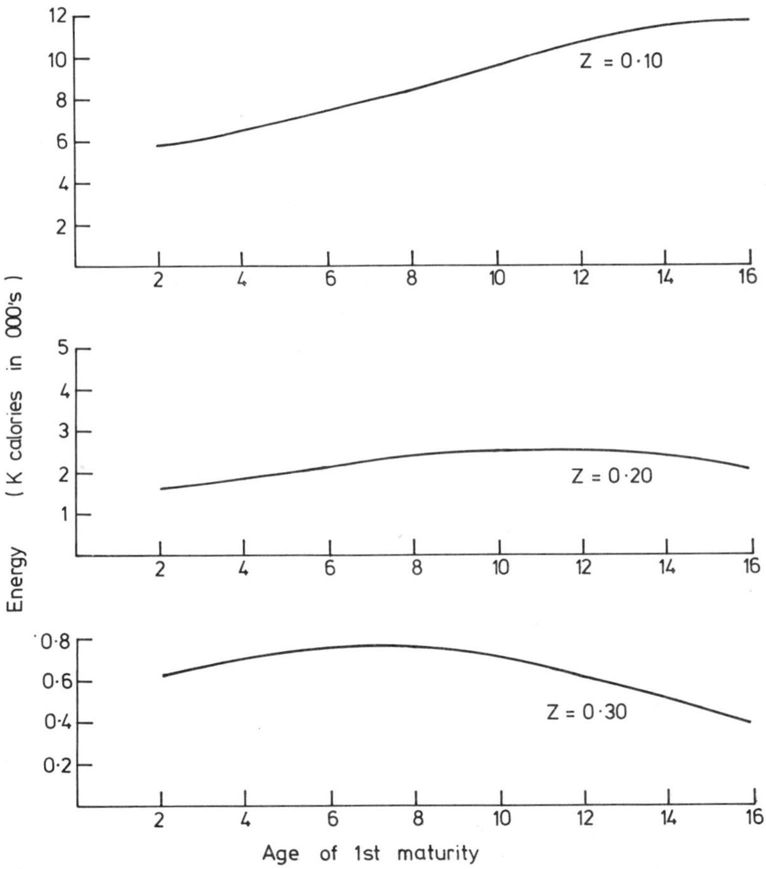

FIG. 13. Relationship between reproductive output per individual and age at first maturity for various values of Z (Arcto-Norwegian cod-type growth).

A consideration that has not been taken into account in these calculations follows from the fact that fish such as cod and haddock derive their energy by exploiting a series of food species. If energy was diverted into reproduction to such an extent that a fish failed to become large enough to reach its final food niche then it would be denied a valuable source of energy. For example there would appear to be little advantage for Arcto-Norwegian cod diverting so much energy into reproduction that it never became large enough to exploit capelin. This would seem to be a good reason therefore for postponing maturity until fish reach at least a certain size.

These results suggest the following conclusions:

(1) A lower limit to the age of first maturity may be influenced not only by the mortality rate but also by the size that has to be attained before a fish can exploit its final food niche efficiently.

(2) The upper limit to the age of first maturity could be determined by the mortality rate.

(3) In general there is no need to assume that an individual fish produces the same relative size of gonad at every spawning. Given more detailed information about the amount of energy going into reproduction at each successive spawning, it ought to be possible to infer something about the way in which natural mortality changes with the age (and/or size) of a fish.

These calculations are based on many assumptions, some of which may be questionable. Nevertheless, it is interesting that the calculated values of the natural mortality rate are consistent with the values of natural mortality currently assumed for the species concerned. It should be noted that no account has been taken of the additional energy required for the spawning migration. This might not be significant for North Sea haddock, but is likely to be important for Arcto-Norwegian cod.

6. Summary

This paper describes attempts to model the growth process in fish and discusses some implications of the results.

In Sections 2 and 3, estimates have been made of the food requirements of various gadoid species (Arcto-Norwegian cod, North Sea haddock and North Sea Norway pout). Food intake has been estimated in terms of the energy required to satisfy basic metabolism, growth rate and reproductive demand of fish of a given body weight. This has led to estimates of the relationship between the rate of food intake and body weight in the different species. It was found that the rate of food intake increased with body weight in cod, haddock and Norway pout. Further, the rate at which food intake increased was found to be different for the different species (greatest for cod and least for Norway pout).

Two points of interest were noted in relation to the increase in rate of food intake with body weight.

(1) Each species tends to exploit a sequence of food organisms throughout

the course of its life. Therefore food energy does not come from a single prey source but from a sequence of prey sources.

(2) A relationship was noted between the rate at which food intake increased with body weight and the form of the growth curve. In general, the greater the rate at which food intake increased with body weight, the more slowly a fish approached its asymptotic body weight (i.e. the smaller the value of the Bertalanffy coefficient K).

Simulation was used to investigate the extreme situation in which the rate of food intake was made constant with body weight. The resultant growth curve was found to approach an asymptotic relationship very rapidly, with a value of K of between 3 and 4.

In Sections 4 and 5, the question is posed "What would happen to growth, if energy, surplus to that needed for metabolism, was used in different ways?" In particular, what would happen to growth if fish produced gonads of a different size, or matured at a different age (or weight)?

Preliminary investigations showed by how much growth should be changed by alterations in these parameters.

The assumption was then made that during the course of evolution, genotypes for all conceivable types of reproductive strategy ought to have been evolved. Existing genotypes should therefore represent the most "successful" out of all possible kinds. As a measure of "success", in this context, it was decided to estimate the total energy going into reproduction throughout the expected life span of an individual. This was calculated for a range of mortality rates for each pattern of food intake and reproductive strategy. The object was to determine the level of mortality at which observed reproductive strategies gave rise to the maximum energy going into reproduction. This mortality, it was argued, ought to be equivalent to the natural mortality rate, provided of course that existing growth rates and reproductive strategies are not significantly different from what they were in the virgin stocks.

These investigations led to two results:

(1) In order to maximize reproductive output using current gonad sizes it was necessary to assume that natural mortality was positively correlated with gonad size (or equivalently positively correlated with the decline in condition factor at spawning time). In this way it was possible to arrive at a generalized set of relationships between mortality rate and gonad size.

(2) In order to maximize reproductive output adopting current ages of first maturity it was concluded that

(a) for North Sea haddock natural mortality should not have been less than 0·2–0·3,

(b) for Arcto-Norwegian cod, natural mortality should not have been more than 0·2–0·3.

Consideration was also given to the way in which fish might conserve energy when food was scarce. Possibilities considered were:

(1) that fish might grow more slowly,

(2) that fish might die at a greater rate than usual,

(3) that less energy might be diverted into reproduction.

It was found that there was sufficient evidence to favour possibility (3) although it is not yet certain to what extent possibilities (1) and (2) also play an important part.

Appendix

Computer simulation

A computer program was written to model growth as given by equation (1).

Each run of the computer program simulated a lifetime of 20 years; each year consisted of 100 cycles. A year was considered to end with spawning and to be divided into two parts. For the first part of the year, cycles 1 to (MG−1), the fish produces no gonads. On cycle MG, the mature fish starts to produce gonads which grow until spawning at the end of the year.

In the following equations relating weights and lengths at the beginning and end of a cycle, W is "weight" of body excluding gonads, Wg is "weight" of gonads and L is length.

Weights are measured in equivalent energy units so that strictly speaking W is the energy content of the somatic body and Wg is the energy content of the gonads. For convenience, however, use will be made of the terms body "weight" and gonad "weight".

For immature fish:

$$W_2 = W_1 + [\lambda\alpha_1 L_1^{r_1} - \alpha_2 W_1^{0\cdot 8} \exp(0\cdot 76V + 0\cdot 081T)]\mathscr{E},$$

where

W_1 = the somatic weight at any moment,

W_2 = the somatic weight $\frac{1}{100}$ year later,

V = swimming speed,

T = temperature,

$\lambda\alpha_1 L_1^{r_1}$ = food energy intake.

α_1, α_2 and r_1 are constants; λ takes account of the differences in summer and winter feeding.

$$\lambda = \begin{cases} 1{\cdot}0 \text{ for cycles 1 to } (MG-1), \\ 0{\cdot}75 \text{ for cycles } MG \text{ to } 100, \end{cases}$$

$$\alpha_2 W_1^{0{\cdot}8} \exp(0{\cdot}76V + 0{\cdot}081T) = \text{maintenance energy},$$

$$\mathscr{E} = 0{\cdot}79 W_1^{-0{\cdot}15} = \text{efficiency of converting surplus energy into body weight.}$$

For mature fish, body weight and gonad weight are interdependent.

$$W_2 = W_1 + [\lambda\alpha_1 L_1^{r_1} - \alpha_2 W_1^{0{\cdot}8} \exp(0{\cdot}76V + 0{\cdot}081T) - \alpha_2 Wg_1^{0{\cdot}8} \exp(0{\cdot}081T)]\mathscr{E} - pWg_1,$$

where

$$\alpha_2 Wg_1^{0{\cdot}8} \exp(0{\cdot}081T) = \text{maintenance energy for developing gonads},$$

$$pWg_1 = \text{incremental weight of gonad produced during each cycle.}$$

For immature fish

$$Wg_1 = Wg_2 = 0,$$

where Wg_1 and Wg_2 are the gonad weight at any two successive cycles. For mature fish

$$Wg_1 = Wg_2 = 0 \quad \text{for cycles 1 to } (MG-1)$$

$$Wg_2 = \begin{cases} \alpha_3 W_1 & \text{for cycle } MG, \\ (1+p)\, Wg_1 & \text{for cycles } (MG+1) \text{ to } 100. \end{cases}$$

For all fish, length at the end of a cycle is given by the equation

$$L_2 = \max\{4{\cdot}8W_2^{\frac{1}{3}}, L_1\}.$$

If Z is the annual mortality rate, the number of survivors N_2 at the end of a cycle is given by

$$N_2 = N_1(1 - 0{\cdot}01Z),$$

where N_1 is the number alive at the beginning of a cycle. Z is constant throughout the lifetime of the fish.

In all the above equations the values of the parameters used in any particular computer run would be defined by the species and stock under study.

References

BAXTER, I. G. and POPE, J. A. (1969). Annual variations in fecundity of Clyde spawning herring, ICES C.M. H:33, 9pp. (mimeo).

BENNETT, D. H. and GIBBONS, J. W. (1975). Reproductive cycles of largemouth bass (*Micropterus salmoides*) in a cooling reservoir. *Trans. Am. Fish. Soc.* **104** (1), 77–82.

BEVERTON, R. J. H. and HOLT, S. J. (1959). A review of the lifespans and mortality rates of fish in nature, and their relation to growth and other physiological characteristics. *Ciba Foundation Symposium on the Lifespan of Animals*, pp. 142–180.

BLAXTER, J. H. S. and HEMPEL, G. (1963). The influence of egg size on herring larvae. *J. Cons. Perm. Int. Explor. Mer* **28**, 211–240.

DAAN, N. (1973). A quantitative analysis of the food intake of North Sea cod, *Gadus morhua. Neth. J. Sea. Res.* **6** (4), 479–517.

DAAN, N. (1975). Consumption and production in North Sea cod, *Gadus morhua*: an assessment of the ecological status of the stock. *Neth. J. Sea Res.* **9** (1), 24–55.

DEELDER, C. L. (1951). A contribution to the knowledge of the stunted growth of perch (*Perca fluviatilis* L.) in Holland. *Hydrobiologia* **3**, 357–378.

FRY, F. E. J. (1949). Statistics of a lake trout fishery. *Biometrics* **5** (1), 22–67.

HISLOP, J. R. G. (1975). The breeding and growth of whiting, *Merlangius merlangus* in captivity. *J. Cons. Int. Explor. Mer* **36** (2), 119–127.

HODDER, V. M. (1965). The possible effects of temperature on the fecundity of Grand Bank haddock. *ICNAF Spec. Publ.* **6**, 515–522.

JONES, R. Competition and co-existence with particular reference to gadoid fish species. ICES Symposium (1975) on the changes in the North Sea Fish Stocks and their causes. *Rapp. P.-V. Réun. Cons. Int. Explor. Mer* (in press).

MACER, C. T. (1974). The reproductive biology of the horse mackerel *Trachurus trachurus* (L.) in the North Sea and English Channel. *J. Fish. Biol.* **6**, 415–432.

MANN, K. H. (1900). Energy transformations by a population of fish in the River Thames. *J. Anim. Ecol.* **34**, 253–275.

MARTIN, N. V. (1970). Long term effects of diet on the biology of the lake trout and the fishery in Lake Opeongo, Ontario. *J. Fish. Res. Bd Can.* **27**, 125–146.

OOSTHUIZEN, E. and DAAN, N. (1974). Egg fecundity and maturity of North Sea cod, *Gadus morhua. Neth. J. Sea Res.* **8** (4), 378–397.

ORTON, J. H. (1929). Reproduction and death in invertebrates and fishes. *Nature*, 3088, **123**, 14–15.

SCOTT, D. P. (1962). Effect of food quantity on fecundity of rainbow trout, *Salmo gairdneri. J. Fish. Res. Bd Can.* **19** (4), 715–731.

TYLER, A. V. and DURN, R. S. (1977). Ration, growth, energy budget and measures of condition in relation to real frequency in winter flounder *Pseudopleuronectes americanus. J. Fish. Res. Bd Can.* (in press).

DE VEEN, J. F. Changes in the North Sea sole stocks. ICES Symposium (1975) on the Changes in the North Sea Fish Stocks, Paper No. 30. *Rapp. P.-V. Réun Cons. Int. Explor. Mer.* (in press).

WEST, W. Q. B. (1970). The spawning biology and fecundity of cod in Scottish waters. PhD thesis, University of Aberdeen.

Winberg, G. G. (1956). "Rate of Metabolism and Food Requirements of Fishes." Nauchnye Trudy Belorusskovo Gosudarstvennovo Universitet a Imens V. I. Lenina, Minsk, 253pp, and *Fish. Res. Bd Can. Trans.* Series No. 194, 202pp.

ZEUTHEN, E. (1953). Oxygen uptake as related to body size in organisms. *Quart. Rev. Biol.* **28** (1), 1–12.

4. Estimation of Fishing Mortality, its Precision and Implications for the Management of Fisheries

J. G. POPE

Fisheries Laboratory, Lowestoft

1. Introduction

The virtual population analysis (VPA) devised by Gulland (1965) is a method of estimating the fishing mortality experienced by year-classes of fish throughout their life. It has proved one of the most valuable ways of obtaining estimates of this parameter, but without accurate fishing effort data it cannot be used confidently for prediction. This paper suggests a new method developed from an idea of Agger *et al.* (1971) which may to some extent overcome this difficulty. However, the method is influenced by the variability of the input data. Clearly this has serious implications for the year-to-year management of fish stocks.

2. Gulland's Virtual Population Analysis: Advantages and Disadvantages

Gulland's virtual population analysis enables fisheries biologists to make estimates of population size and fishing mortality (F) at each age independently of measures of fishing effort. Since fishing effort data are often not proportional to F this is a very real advantage and is of great value in elucidating the structure of a fishery. A brief description of the method can be found in Pope (1972). Its chief disadvantage stems from the fact that $n+1$ independent estimates (viz. the population at the age of first capture and the fishing mortality at each of the n ages of the year-class) have to be obtained from n equations based on the catch at age data from the n ages of the year-class. Clearly it is not possible to solve these equations without making some assumptions about at least one of the parameters. Gulland's solution was to estimate or guess the fishing mortality in the most recent year and then to calculate the consequent populations and fishing mortalities in each of the preceding years. It has been shown by Pope (1972) and by

Agger *et al.* (1971) that this assumption of the fishing mortality of the oldest age, F_t, causes progressively smaller errors on the estimates of fishing mortality and population size calculated for the younger ages of the year-class. Thus Gulland's solution is extremely valuable in understanding a fishery in a historic sense and for elucidating its population dynamics. However, as the oldest age of each year-class of a fish stock is necessarily the most recent, it follows that the estimate of the current population of fish is only as good as the current estimate or guess (usually guess) of the fishing mortality in the last year. While it is true that this assumption may be made more objective by making use of any fishing effort data available, the accuracy of these is rarely sufficient to give a very clear view of the current situation except in those fisheries that have a very constant exploitation rate. The relation of these problems to the setting of catch quotas has been discussed by Pope and Garrod (1973). It is clear that alternative methods of estimating fishing mortality and population size are urgently required.

3. A Possible Alternative to Virtual Population Analysis

One attempt at developing an alternative to Gulland's method was made by Agger *et al.* (1971) who suggested a least-squares solution which estimated the values of population at each age, the fishing mortality in each year and the natural mortality from the catch at age/year matrix. The author understands that this solution attempted to estimate these parameters for the fully recruited ages only and was based on the assumption that the error in catches was approximately normally distributed. Subsequent usage by the ICES North Sea Herring Assessments Working Group revealed problems. However, despite these the basic idea of using a nonlinear least-squares estimate of fishing mortality and population is one that is extremely attractive from a statistical standpoint and deserves close attention. It is therefore worth considering what problems were associated with the proposed method and seeing if they can be overcome. For example, consider the catch at age matrix given in Table 1(a). If the natural mortality is assumed to be 0·2, the fishing mortalities at each year and age given in Table 1(b) will, apart from rounding errors, satisfy the catch data, but equally so will the different fishing mortalities given in Table 1(c). In fact there are infinitely many possible solutions to this problem, because catch data from the fully exploited ages of a year-class form a sequence which is identical to that of the adjacent year-classes, except for random variations and a constant multiplier, the successive terms being displaced by one year. This suggests that a least-squares solution would not be unique in this case and also that this problem might be overcome by considering ages of fish which are only partially recruited to the fishery.

Another part of the original model which is worth considering is the assumption that errors in catch samples are normal. Gulland (1955) suggests that the best form of age sampling would be one in which the numbers caught at each age had approximately the same coefficient of variation. While practical sampling schemes rarely achieve this ideal, as an assumption

TABLE 1. An example of a non-unique solution

	Age			
Year	1	2	3	4
(a) Catch data from a homogeneous fish stock with $M = 0{\cdot}2$				
1960	154	77	22	52
1961	108	97	48	14
1962	76	63	56	28
1963	61	69	57	51
1964	169	52	58	48
(b) Possible values of coefficient of fishing mortality				
1960	0·23	0·23	0·23	0·23
1961	0·22	0·22	0·22	0·22
1962	0·20	0·20	0·20	0·20
1963	0·27	0·27	0·27	0·27
1964	0·39	0·39	0·39	0·39
(c) Other possible values of coefficient of fishing mortality				
1960	0·25	0·25	0·25	0·25
1961	0·25	0·25	0·25	0·25
1962	0·22	0·22	0·22	0·22
1963	0·32	0·32	0·32	0·32
1964	0·49	0·49	0·49	0·49

it is probably better than the alternative that each catch has an equivalent variance. This suggests that a method based on a least-squares solution of a logarithmic transformation might be more appropriate.

Bearing in mind the points raised in the last two paragraphs, a new least-squares model can be developed. For example, consider the catch ${}_aC_y$ generated by a fishing fleet on a homogeneous fish stock. Assume that in year y the fleet generates a fishing mortality of F_n on fully recruited age groups and that this is then modified on the various ages by partial recruitment factors S_a, then fishing mortality at age a in year y will be $F_y . S_a$. If the natural mortality is M the relation of ${}_aC_y$ to ${}_aP_y$, the population of age a, at the beginning of year y, will be

$$ {}_aC_y . \varepsilon = {}_aP_y . F_y . S_a \{1 - \exp[-(F_y . S_a + M)]\} / (F_y . S_a + M). \tag{1} $$

ε in this equation is the sampling error and is assumed to be log normal. Apart from ε, equation (1) is the usual model assumed for most fisheries based on a mixture of partially and fully recruited fish. In practice ${}_aC_y$ is the only variable which can be directly measured and the matrix of ${}_aC_y$ for all a and all y forms the input. To stabilize the sampling variance a $\log_e$ (ln) transformation is made which gives

$$ \begin{aligned} \ln({}_aC_y) + \varepsilon' = \ln({}_aP_y) + \ln(F_y . S_a) - \ln(F_y . S_a + M) \\ + \ln\{1 - \exp[-(F_y . S_a + M)]\}, \end{aligned} \tag{2} $$

where ε' is the transformed error term and is normally distributed $n(0, \sigma^2)$ and M is the instantaneous coefficient of natural mortality. The problem is to find values of ${}_aP_y$, F_y and S_a which satisfy the catch matrix in some sense. It can be simplified by eliminating ${}_aP_y$ by subtracting the next diagonal term from each form of equation (2) (i.e. subtracting equation (2) with $a = i+1$, $y = j+1$ from equation (2) with $a = i$, $n = j$). This yields

$$ \begin{aligned} \ln({}_aC_y / {}_{a+1}C_{y+1}) = {} & \ln({}_aP_y / {}_{a+1}P_{y+1}) + \ln(F_y . S_a / F_{y+1} . S_{a+1}) \\ & - \ln(F_y . S_a + M) + \ln(F_{y+1} . S_{a+1} + M) \\ & + \ln\{1 - \exp[-(F_y . S_a + M)]\} \\ & - \ln\{1 - \exp[-(F_{y+1} . S_{a+1} + M)]\} + \varepsilon'', \end{aligned} \tag{3} $$

where ε'' is $n(0, 2\sigma^2)$. Since

$$ \ln({}_aP_y / {}_{a+1}P_{y+1}) = F_y . S_a + M, \tag{4} $$

it follows that equation (3) can be reduced to a form which contains only ${}_aC_y$, F_y, S_a, M and ε''.

For a least-squares solution the problem then becomes one of finding the values of F_y and S_a which minimize $[\ln({}_aC_y' / {}_{a+1}C_{y+1}') - \ln({}_aC_y / {}_{a+1}C_{y+1})]^2$, where ${}_aC_y$ refers to observed values of the catch at age data and ${}_aC_y'$ to the values of catch ratio calculated from the values of F_y and S_a by the combination of equations (3) and (4). This is made under the assumption that M is known.

The minimization is achieved by a modified form of the steepest-descent method but other minimization methods could also be used. Doubleday (1975) gives an alternative minimization algorithm.

4. Testing the Model

The most obvious test of this model is to solve F_y and S_a from catch data generated from an imaginary stock with known $_aP_y$, F_y, S_a and M. If the catch is generated with no sampling error, the proposed method, if it is satisfactory, should be expected to output the values of F_y and S_a which were used to generate the catch data. Table 2 shows the results of one such

TABLE 2. Comparison of the values of fishing mortality and selection coefficients used to generate catches from a simulated population and the corresponding values estimated by the proposed method

Fishing mortality

	Year									
	1	2	3	4	5	6	7	8	9	10
True value	0·23	0·22	0·20	0·27	0·39	0·43	0·45	0·46	0·48	0·41
Value estimated by least squares	0·23	0·22	0·19	0·27	0·39	0·43	0·44	0·46	0·48	0·40

Selection coefficients

	Age							
	1	2	3	4	5	6	7	8
True value	0·01	0·08	0·41	0·67	1·00	1·00	1·00	1·00
Value estimated by least squares	0·01	0·08	0·41	0·67	1·00	1·00	1·00	1·00

test. Slight differences occur because of rounding errors, but it is clear that the method has effectively reproduced the values of the F_y and S_a.

The next form of test is similar to the last, but with the generated catch data moderated by a series of log normal random numbers. The results from such catch data should not, of course, be expected to reproduce the true values of F_y and S_a, but a series of runs with the same basic data and with differing random numbers should be expected to generate average answers which do not differ systematically from the true ones. The method should produce unbiased estimates of F_y and S_a. Tables 3, 4, 5 and 6 show

TABLE 3. Comparison of the average fishing mortality ($\bar{F}$), from 10 runs of the model with catch data having a coefficient of variation of 5%, with the true values of fishing mortality (F_t) for each age/year

Age		Year 1	2	3	4	5	6	7	8	9	10
1	$\bar{F}$	0·002	0·002	0·002	0·003	0·004	0·004	0·004	0·004	0·004	0·003
	F_t	0·002	0·002	0·002	0·003	0·004	0·004	0·005	0·005	0·005	0·004
2	$\bar{F}$	0·02	0·02	0·02	0·02	0·03	0·03	0·03	0·03	0·03	0·03
	F_t	0·02	0·02	0·02	0·02	0·03	0·03	0·04	0·04	0·04	0·03
3	$\bar{F}$	0·10	0·10	0·08	0·11	0·16	0·17	0·17	0·17	0·17	0·13
	F_t	0·09	0·09	0·08	0·11	0·16	0·18	0·18	0·19	0·20	0·17
4	$\bar{F}$	0·17	0·16	0·14	0·19	0·27	0·29	0·29	0·29	0·29	0·23
	F_t	0·15	0·15	0·13	0·18	0·26	0·29	0·30	0·31	0·32	0·27
5	$\bar{F}$	0·26	0·24	0·21	0·28	0·41	0·44	0·44	0·44	0·44	0·34
	F_t	0·23	0·22	0·20	0·27	0·39	0·43	0·45	0·46	0·48	0·41
6	$\bar{F}$	0·27	0·26	0·22	0·30	0·43	0·47	0·47	0·47	0·47	0·36
	F_t	0·23	0·22	0·20	0·27	0·39	0·43	0·45	0·46	0·48	0·41
7	$\bar{F}$	0·29	0·27	0·23	0·32	0·46	0·49	0·49	0·50	0·50	0·38
	F_t	0·23	0·22	0·20	0·27	0·39	0·43	0·45	0·46	0·48	0·41
8	$\bar{F}$	0·31	0·30	0·26	0·35	0·50	0·54	0·54	0·54	0·54	0·42
	F_t	0·23	0·22	0·20	0·27	0·39	0·43	0·45	0·46	0·48	0·41

TABLE 4. Comparison of the average fishing mortality ($\bar{F}$), from 10 runs of the model with catch data having a coefficient of variation of 10%, with the true values of fishing mortality (F_t) for each age/year

Age		Year 1	2	3	4	5	6	7	8	9	10
1	$\bar{F}$	0·002	0·002	0·002	0·003	0·004	0·004	0·004	0·004	0·004	0·003
	F_t	0·002	0·002	0·002	0·003	0·004	0·004	0·005	0·005	0·005	0·004
2	$\bar{F}$	0·02	0·02	0·02	0·02	0·03	0·03	0·03	0·03	0·03	0·02
	F_t	0·02	0·02	0·02	0·02	0·03	0·03	0·04	0·04	0·04	0·03
3	$\bar{F}$	0·10	0·09	0·08	0·11	0·16	0·17	0·17	0·17	0·17	0·13
	F_t	0·09	0·09	0·08	0·11	0·16	0·18	0·18	0·19	0·20	0·17
4	$\bar{F}$	0·16	0·16	0·13	0·18	0·26	0·28	0·28	0·28	0·29	0·22
	F_t	0·15	0·15	0·13	0·18	0·26	0·29	0·30	0·31	0·32	0·27
5	$\bar{F}$	0·25	0·24	0·20	0·28	0·40	0·42	0·42	0·43	0·43	0·32
	F_t	0·23	0·22	0·20	0·27	0·39	0·43	0·45	0·46	0·48	0·41
6	$\bar{F}$	0·27	0·25	0·22	0·30	0·43	0·46	0·46	0·46	0·46	0·35
	F_t	0·23	0·22	0·20	0·27	0·39	0·43	0·45	0·46	0·48	0·41
7	$\bar{F}$	0·28	0·26	0·23	0·31	0·44	0·47	0·48	0·48	0·48	0·36
	F_t	0·23	0·22	0·20	0·27	0·39	0·43	0·45	0·46	0·48	0·41
8	$\bar{F}$	0·30	0·29	0·25	0·34	0·48	0·51	0·51	0·52	0·52	0·39
	F_t	0·23	0·22	0·20	0·27	0·39	0·43	0·45	0·46	0·48	0·41

TABLE 5. Comparison of the average fishing mortality ($\bar{F}$), from 10 runs of the model with catch data having a coefficient of variation of 20%, with the true values of fishing mortality (F_t) for each age/year

Age		Year 1	2	3	4	5	6	7	8	9	10
1	$\bar{F}$	0·002	0·002	0·002	0·002	0·004	0·004	0·004	0·004	0·004	0·003
	F_t	0·002	0·002	0·002	0·003	0·004	0·004	0·005	0·005	0·005	0·004
2	$\bar{F}$	0·02	0·02	0·02	0·02	0·03	0·03	0·03	0·03	0·03	0·02
	F_t	0·02	0·02	0·02	0·02	0·03	0·03	0·04	0·04	0·04	0·03
3	$\bar{F}$	0·10	0·09	0·08	0·11	0·15	0·16	0·16	0·16	0·16	0·12
	F_t	0·09	0·09	0·08	0·11	0·16	0·18	0·18	0·19	0·20	0·17
4	$\bar{F}$	0·16	0·15	0·13	0·18	0·26	0·28	0·28	0·27	0·28	0·21
	F_t	0·15	0·15	0·13	0·18	0·26	0·29	0·30	0·31	0·32	0·27
5	$\bar{F}$	0·24	0·23	0·20	0·27	0·39	0·40	0·40	0·40	0·41	0·29
	F_t	0·23	0·22	0·20	0·27	0·39	0·43	0·45	0·46	0·48	0·41
6	$\bar{F}$	0·27	0·26	0·22	0·30	0·44	0·46	0·46	0·45	0·45	0·33
	F_t	0·23	0·22	0·20	0·27	0·39	0·43	0·45	0·46	0·48	0·41
7	$\bar{F}$	0·29	0·27	0·23	0·32	0·46	0·48	0·47	0·47	0·48	0·34
	F_t	0·23	0·22	0·20	0·27	0·39	0·43	0·45	0·46	0·48	0·41
8	$\bar{F}$	0·31	0·30	0·25	0·34	0·50	0·52	0·51	0·50	0·51	0·36
	F_t	0·23	0·22	0·20	0·27	0·39	0·43	0·45	0·46	0·48	0·41

TABLE 6. Comparison of the average fishing mortality ($\bar{F}$), from 10 runs of the model with catch data having a coefficient of variation of 40%, with the true values of fishing mortality (F_t) for each age/year

Age		Year 1	2	3	4	5	6	7	8	9	10
1	$\bar{F}$	0·002	0·002	0·002	0·002	0·004	0·004	0·004	0·004	0·004	0·003
	F_t	0·002	0·002	0·002	0·003	0·004	0·004	0·005	0·005	0·005	0·004
2	$\bar{F}$	0·02	0·02	0·01	0·02	0·03	0·03	0·03	0·03	0·03	0·02
	F_t	0·02	0·02	0·02	0·02	0·03	0·03	0·04	0·04	0·04	0·03
3	$\bar{F}$	0·09	0·09	0·08	0·10	0·15	0·15	0·15	0·14	0·15	0·11
	F_t	0·09	0·09	0·08	0·11	0·16	0·18	0·18	0·19	0·20	0·17
4	$\bar{F}$	0·16	0·15	0·13	0·18	0·26	0·26	0·27	0·25	0·28	0·21
	F_t	0·15	0·15	0·13	0·18	0·26	0·29	0·30	0·31	0·32	0·27
5	$\bar{F}$	0·23	0·22	0·18	0·25	0·38	0·37	0·37	0·36	0·39	0·27
	F_t	0·23	0·22	0·20	0·27	0·39	0·43	0·45	0·46	0·48	0·41
6	$\bar{F}$	0·29	0·27	0·23	0·31	0·47	0·46	0·46	0·45	0·47	0·32
	F_t	0·23	0·22	0·20	0·27	0·39	0·43	0·45	0·46	0·48	0·41
7	$\bar{F}$	0·30	0·28	0·24	0·33	0·51	0·49	0·50	0·48	0·53	0·37
	F_t	0·23	0·22	0·20	0·27	0·39	0·43	0·45	0·46	0·48	0·41
8	$\bar{F}$	0·34	0·32	0·27	0·37	0·56	0·55	0·56	0·53	0·59	0·41
	F_t	0·23	0·22	0·20	0·27	0·39	0·43	0·45	0·46	0·48	0·41

the results of such comparisons made in this form, in which the random numbers were chosen to give the catch data coefficients of variation of 5, 10, 20 and 40%. It was only possible to make 10 runs at each level in the time available, but it can be seen from the tables that the average of those 10 runs ($\bar{F}$) was close to the true value (F_t): this result is encouraging. The F in the last (tenth) year and the F of the last age (eighth) seem to deviate most from the true value. The coefficient of variation (Tables 7, 8, 9 and 10)

TABLE 7. Coefficients of variation of fishing mortality obtained from 10 runs of the model with catch data having a coefficient of variation of 5%. Results are for the fully recruited ages only for each year

	Year									
Age	1	2	3	4	5	6	7	8	9	10
5	6	4	5	3	4	4	5	6	10	12
6	6	5	6	5	5	5	5	7	9	12
7	8	7	6	5	7	6	6	8	10	12
8	11	10	10	8	9	9	8	8	11	12

TABLE 8. Coefficients of variation of fishing mortality obtained from 10 runs of the model with catch data having a coefficient of variation of 10%. Results are for the fully recruited ages only for each year

	Year									
Age	1	2	3	4	5	6	7	8	9	10
5	7	7	5	6	8	6	7	10	15	18
6	11	11	10	9	10	8	7	12	14	17
7	12	12	10	11	13	10	10	14	17	20
8	17	18	16	16	16	15	13	17	17	18

of the fully recruited fishing mortalities obtained suggests that the accuracy of a realization is related to the accuracy of the catch data, the precision of the final year's fishing mortalities being approximately half that of the catch data. This suggests that the method's use should be restricted to well sampled catch data.

The value of the sum of squares is presumably an indication of the degree of variability to the catch data. For example, the average least sum of squares obtained from the 10 runs shown in Table 4 was 0·9482. For the array of 8 ages and 10 years there were $(m-1)(n-1)-n-m+1$ degrees of freedom, that is, 46 degrees of freedom. Hence the means square of residuals was 0·02061, and this divided by 2 (see equation (3)) gives an estimate of the variance of $\log_e$ transformed catch data as 0·01030. This would seem consistent with the 0·01 which would be expected from catch data with a 10% coefficient of variation.

TABLE 9. Coefficients of variation of fishing mortality obtained from 10 runs of the model with catch data having a coefficient of variation of 20%. Results are for the fully recruited ages only for each year

	Year									
Age	1	2	3	4	5	6	7	8	9	10
5	13	11	5	9	13	11	15	21	32	44
6	18	15	11	11	14	12	11	20	27	37
7	24	21	14	19	21	18	18	23	34	45
8	28	26	19	22	22	22	16	19	28	39

TABLE 10. Coefficients of variation of fishing mortality obtained from 10 runs of the model with catch data having a coefficient of variation of 40%. Results are for the fully recruited ages only for each year

	Year									
Age	1	2	3	4	5	6	7	8	9	10
3	11	20	19	14	23	16	27	37	52	74
4	17	19	25	25	31	31	45	48	70	97
5	19	15	12	17	28	20	32	43	60	87
6	24	20	19	18	29	22	25	43	50	70
7	30	26	22	31	44	30	45	55	73	101
8	26	24	22	31	42	29	47	56	76	107

5. Consideration of the Uniqueness of Solutions by this Model

One of the obvious doubts associated with this technique is the large number of parameters that need to be fitted to a comparatively small data set. For example, in the previous examples of the model, 10 fishing mortalities and 7 partial recruitment factors were estimated (the eighth partial recruitment factor may be set arbitrarily). These parameters have to be fitted to 63 data points $[(10-1).(8-1)]$ which gives only a little more than three data points per parameter. Clearly if the problem were viewed as a regression this would be a very unsatisfactory data to parameter ratio. One way of improving this would be to find functions which describe the fishing mortality and partial recruitment with less parameters. For example, the partial recruitment at age a $[S(a)]$ might be approximated by

$$S(a) = 1-\exp(Ba),$$

where B was some parameter. This would help the data to parameter ratio, but it is difficult to see how similar approximations could usefully be applied to the fishing mortality.

Fortunately, the problem may be viewed in a different way and this in turn sheds considerable light on the nature of this model. The following approximation

$$\frac{1-\exp(-Z)}{Z} \simeq \exp\left(-\frac{Z}{2}+\frac{Z^2}{24}\right) \tag{5}$$

enables equation (3) to be rewritten approximately as

$$\ln({}_aC_y/{}_{a+1}C_{y+1}) = \ln(F_y/F_{y+1})+\ln(S_a/S_{a+1}) + \frac{F_y.S_a}{2}+\frac{F_{y+1}.S_{a+1}}{2}+M + \frac{(F_y.S_a+M)^2}{24}-\frac{(F_{y+1}.S_{a+1}+M)^2}{24}+\varepsilon'' \tag{6}$$

It is clear from equation (6) that the first two terms are simply year and age effects while the third and fourth terms are part year and age effects and part some interaction term. The fourth term is a constant M and the sixth and seventh terms are generally small and of opposite sign and therefore unlikely to affect the answer very much. Thus the model has some of the features of a two-way analysis of variance with no replicates.

How closely could the transformed data be fitted by a two-way analysis of variance? In order to investigate this question a two-way analysis of variance was made of the original exact test data used in Section 4. The analysis of variance which results is shown in Table 11. In this example, the year and age effects account for the major part of the variance. Since the data had no variance with respect to the model described in equation (3) the residual term of the two-way analysis of variance is that fraction of the variance which the former model is required to explain, that is, the interaction terms. In this example this amounts to only 0·2% of the total variance thus suggesting, as does equation (6), the close correspondence of the two models.

In one sense this is encouraging. A simple two-way analysis of variance can be performed on the transformed catch data to indicate whether or not there are differences between the various year effects and the various age effects. The number of parameters which the mortality model uses to explain the observations is almost the same as in the case of an analysis of variance. Fitting this number of terms thus seems almost respectable. However, while the analysis of variance model would have $n-1$ year effects (where n is the number of years) the model proposed here has n. It can be shown that if

TABLE 11. Analysis of variance for the exact test data

Source	D.F.	Sum of squares
Ages	6	53·067
Years	8	2·176
Interactions	48	0·092
Total	62	55·335

fishing mortality in the last year (*Fn*) is chosen arbitrarily, values of all the other fishing mortalities and partial recruitment factors can be chosen to give the correct row and column totals. Thus the only basis for choosing one particular *Fn* rather than another must be how well it and the other parameters which result from it fit the interaction terms of equation (6). From the example it would seem likely that the increase in the proportion of the variance explained as a result of getting *Fn* accurately fitted will be rather small. Consequently *Fn*, the fishing mortality in the last year, is unlikely to be fitted with very great precision. This is, of course, in line with the results of the Monte Carlo type analysis performed in Section 4.

How much of the data sum of squares is available in the interaction terms of equation (6) is dependent on the configuration of the problem. The more that the fishing mortality changes through time and the greater the differences in the partial recruitment factors, the greater the amount of evidence in the interaction terms is likely to be and hence the greater the precision with which the final term can be fitted. It is necessary for both sets of parameters to vary for there to be any evidence. This is best demonstrated by the following examples. Table 12 gives the sum of squares due to years, age and interaction terms for three sets of data. Data set 1 had a constant fishing mortality for all years and partial recruitment was constant for all ages. Data set 2 had a variable fishing mortality with time but a constant partial recruitment factor. Data set 3 had both variable fishing mortality with time and variable partial recruitment factors with age. It is clear from this analysis that it is only when both fishing mortality and partial recruitment vary that the model suggested in equation (3) fits the data any more closely than a two-way analysis of variance. It is therefore only in these circumstances that the final parameter could possibly be fitted uniquely.

In practice, of course, it is not possible to examine the residual terms of the two-way analysis of variance for the existence of this evidence since normally the residual term contains not only the interaction terms but also the error. This problem can, however, be circumvented by choosing the fishing mortality in the final year and then fitting the other parameters so as to minimize the sum of squares function. This can be done for a range of *Fn* values (the value of fishing mortality in the final year) and the residual sum of squares of the model can be computed. If the evidence exists which will fix *Fn*, the resulting graph of the sum of squares function on *Fn* will have a single minimum at the "best" value of *Fn*, that is, the value of *Fn* which gives, together with the consequential values of the other fishing mortalities and partial recruitment factors, the least-squares fit to the data. Figure 1 shows the graph generated by fitting the test data discussed in Section 4 in this fashion. In this case there is a clear minimum and *Fn* can be fitted. By contrast a similar analysis of the North Sea cod showed a large

TABLE 12. Analyses of variance for data sets 1, 2 and 3

		Sum of squares		
	D.F.	Data set 1	Data set 2	Data set 3
Rows	6	0·0	0·0	64·9757
Columns	9	0·0	12·852	7·9056
Interactions	54	0·0	0·0	0·72759
Total	69	0·0	12·8522	73·608914

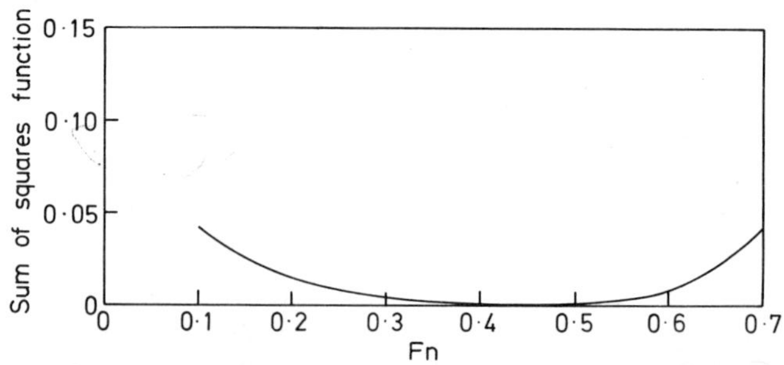

FIG. 1. The sum of squares function relative to various assumed values of fishing mortality in the last year (*Fn*) for the artificial data used to test the model in Section 4.

amount of residual variance and no worth-while minimum could be detected and this technique therefore cannot be satisfactorily used for this stock to estimate *Fn*. The method does, however, produce plausible solutions subject to the assumption about *Fn*. Thus, in this example this technique of fitting the parameters, subject to the final year's fishing mortality being fixed, could be regarded as an alternative to VPA and having the advantage over VPA that only one fishing mortality needs to be estimated or guessed instead of a separate estimate for the oldest age of each year-class. The fact that *Fn* cannot be fitted to the North Sea cod is not particularly surprising since there was in fact little change in the partial recruitment factor after age 2.

6. Discussion

Fisheries management at present makes considerable use of annual catch quotas to regulate fisheries, seeking to achieve a management objective (frequently expressed in terms of fishing mortality) by limiting the total weight of fish caught. In order to estimate an annual catch quota it is necessary to estimate the population size of the incoming year-class and the populations of those ages of fish which were in the fishery in the last year for which catch-at-age data are available. The effects of variance on these various estimates and consequently the precision with which the objective is achieved are discussed by Pope and Garrod (1973). It is sufficient here to point out that the catch quota estimated is frequently considerably in error and that in many cases this is a result of errors in estimates of the population for the year-classes for which catch at age data are available. These errors result from errors in estimating the fishing mortality in the final year for which catch-at-age data are available, since if catch and fishing mortality are known the populations can be calculated.

The problem thus is to estimate as accurately as possible, *Fn*, the fishing mortality in the final year and the partial recruitment factors, which enable it to be applied to the varied ages of fish. VPA does not directly estimate *Fn* and the usual means of estimating it is by regressing VPA fishing mortalities from past years against some measure of fishing effort. The fishing effort is then used to estimate *Fn*. This technique is subject to considerable errors since effort data are always variable and often ambiguous. Clearly the Section 3 model might be useful since in some circumstances it can estimate *Fn*. On any real life data, however, this estimate is likely to be rather variable, and the method should be applied with due caution. The degree of variability can be judged from the examples of Section 4 and would appear to be related to the degree of variability of the catch at age data. It would also appear from Section 5 that an estimate of *Fn* can only be made using this model when fishing mortality changes from year to year and the partial recruitment factors change from age to age.

Given the degree of variation in *Fn* which will occur howsoever it is estimated, it must follow that catch quotas will in general be in error and thus the management objective will not in fact be achieved precisely. The stock of fish will in some years be overexploited and in other years underexploited (assuming the estimate is not permanently biased). For some stocks of fish this may not be of any great consequence because the results may balance out, but for some stocks, perhaps with a strong stock and recruitment relationship, this variation in the achievement of the objective could lead to disaster. Consequently, it is possible that considerations of the precision with which a management optimum can be met may lead to the choice of objectives which would be less optimal were they precisely achieved, but which are less dangerous given the degree of variation met with in practice. For stocks of fish for which catch quota estimates are likely to be inaccurate it would be worth while considering managing the stock at some level of fishing mortality below the F_{max} level.

References

AGGER, P., BOETIUS, I. and LASSEN, H. (1971). On errors in the virtual population analysis. ICES C.M. 1973/H:16 (mimeo).

DOUBLEDAY, W. G. (1975). A least squares approach to analyzing catch at age data. *Ann. Meet. Int. Commn NW. Atlant. Fish., Res. Doc. No.* 35, *Serial No.* 3514.

GULLAND, J. A. (1955). Estimation of growth and mortality in commercial fish populations. *Fishery Invest. Lond.*, Ser. 2, **18** (9).

GULLAND, J. A. (1965). Estimation of mortality rates. Annex to Arctic Fisheries Working Group Report (meeting in Hamburg, January 1965). ICES C.M. 1965, Doc. No. 3 (mimeo).

POPE, J. G. (1972). An investigation of the accuracy of virtual population analysis using cohort analysis. *Res. Bull. Int. Commn NW. Atlant. Fish.* **9**, 65–74.

POPE, J. G. and GARROD, D. J. (1973). A contribution to the discussion of the effects of error on the action of catch quotas and effort quotas. *Ann. Meet. Int. Commn NW. Atlant. Fish.*, 1973, *Res. Doc. No.* 110, *Serial No.* 3074.

5. The Estimation of Mortality from Capture–Recapture Experiments

M. D. Nicholson and J. A. Pope

Marine Laboratory, Aberdeen

1. Introduction

The use of tagging techniques in the study of animal populations is widespread particularly in fisheries research. Early applications of the method in this field were mainly directed towards discussing patterns of movement and migration and providing a basis for estimating total population size. During the past quarter of a century tagging experiments have been extensively used to provide estimates of the mortality rates caused by the commercial exploitation of marine fish populations as a necessary preliminary for the establishment of rational resource management.

This has stimulated considerable research into the development of efficient experimental procedures and associated statistical techniques of estimation. An excellent review of the statistical aspects of capture–recapture experiments has been given by Cormack (1968). Techniques of special relevance in fisheries research have been reviewed by Pope (1961) and Jones (1966).

Whilst considerable attention has been devoted to the derivation of estimates of population parameters such as mortality rates, far less attention has been given to the sampling properties of such estimates. This paper forms part of a study of the robustness of estimates and inferential procedures from data obtained from tagging experiments.

An account of basic statistical procedures would be out of place here but it is useful to define two concepts central to the present study. We take statistical inference to mean a process whereby information (data) is used to provide, through a probability model, a description of a real situation. A robust procedure is one which is not significantly affected by moderate departures from the assumptions on which its validity has been derived.

This work sprang largely from the authors' involvement with the analysis of recaptures from a large-scale tagging experiment in the North Sea (Report of the ICES Working Group, 1975). This experiment had been undertaken to provide estimates of fishing mortality for immature herring in the Bløden industrial herring fishery which is located in the south-eastern part of the North Sea.

The experiment was carried out over the period July 1969 to March 1970, during which time approximately 57 000 tagged herring were liberated. Small internal metal tags were used which were recovered by electromagnets when fish were being processed into fish meal. The fish were released in batches of approximately 2000. For the specific purpose of the present study data from a series of four such batches, from which a total of 6539 tagged fish were effectively released during the period 31 July to 7 August 1969, will be used. These data are given in Table 1.

TABLE 1. Recaptures by month at Esbjerg, Denmark from 6539 tagged fish effectively liberated into the south-eastern North Sea over the period 31 July–7 August 1969
Source: Report ICES Working Group 1975 table 36(a)

Month	Number recaptured	Month	Number recaptured
August 1969	359	July 1970	7
September	66	August	16
October	33	September	10
November	11	October	6
December	0	November	1
January 1970	4	December	0
February	133	January 1971	0
March	7	February	5
April	14	March	0
May	18	April	3
June	0		

2. A Fixed Parameter Model

The fundamental assumption in the models discussed here is that the number, $N(t)$, of animals alive at time t in a population subject to both fishing and natural mortality, acting independently, obeys the relationship

$$\frac{dN(t)}{dt} = -[F(t)+M(t)]\,N(t), \qquad (1)$$

where $F(t)$ and $M(t)$ are coefficients of instantaneous fishing and natural mortality, respectively. Natural mortality may be considered to include all losses from the population other than those due to fishing.

If $F(t) = F$ and $M(t) = M$, where F and M are constants, then (1) has the solution

$$N(t) = N(0)\exp(-Zt), \tag{2}$$

where $Z = F+M$, and the number of animals caught in the interval (t_1, t_2) is

$$FN(0)\exp(-Zt_1)\,[1-\exp(t_2-t_1)]/Z. \tag{3}$$

Writing N in place of $N(0)$ and letting t_2 tend to infinity, the probability that n individuals will subsequently be recaptured is given by

$$P(n/N) = \binom{N}{n}(F/Z)^n(M/Z)^{N-n}. \tag{4}$$

If the recaptures occur at times $t_1, t_2, \ldots, t_n$, assumed exactly known, each will have a probability density function

$$f(t)\,dt = Z\exp(-Zt)\,dt, \quad 0 \leqslant t \leqslant \infty. \tag{5}$$

Chapman (1961) has shown that for this model almost unbiased estimates of F and M are given by

$$\hat{F}_c = n(n-1)/N\sum t_i \qquad n \geqslant 2, \tag{6a}$$

$$\hat{M}_c = (N-n)(n-1)/N\sum t_i, \quad n \geqslant 2. \tag{6b}$$

These expressions are similar to the approximate maximum-likelihood estimates given by Gulland (1955). Approximate expressions, of the order $1/N$, for the variances of $\hat{F}_c$ and $\hat{M}_c$ and their covariance are

$$\operatorname{var}(\hat{F}_c) \doteqdot \frac{F(F+2M)}{N} \tag{7a}$$

$$\operatorname{var}(\hat{M}_c) \doteqdot \frac{M(F^2+FM+M^2)}{FN} \tag{7b}$$

$$\operatorname{var}(\hat{F}_c, \hat{M}_c) \doteqdot \frac{M^2}{N}. \tag{7c}$$

Chapman (1961) derived the expression

$$\operatorname{var}(\hat{F}_c) = \frac{F(F+2M)}{N} + \frac{2(F+M)}{N^2}. \tag{7d}$$

The obvious approach to setting limits to the estimates (6) would be to use (7a–c) to define a confidence ellipse corresponding to, say, a probability level of 0·95.

In practice it will be desirable to base estimates on the number of recaptures which have occurred within a certain finite period of time (T) from the commencement of the experiment. In such a situation Paulik (1963) has suggested reparameterization of the model in terms of

$$\mu = F[1-\exp(-ZT)]/Z, \tag{8a}$$

$$s = \exp(-Z). \tag{8b}$$

The parameters (μ, s) have the desirable property of independence, i.e. $\operatorname{cov}(\hat{\mu}, \hat{s}) = 0$. Paulik suggested that approximate confidence intervals for μ and s might be obtained by treating $\hat{\mu}$ and $\hat{s}$ as Normally distributed with variances given by

$$\operatorname{var}(\hat{\mu}) = \mu(1-\mu)/N, \tag{9a}$$

$$\operatorname{var}(\hat{s}) \doteqdot (s \ln s)^2/N\mu. \tag{9b}$$

Whilst this has obvious statistical advantages the derivation of estimates of F and M and their confidence ellipse from (8) and (9) is not straightforward.

In comparison with this approach, the ease of computation and the directness with which the variance and covariance expressions for Chapman's estimates may be obtained make them much more attractive provided they are robust to truncating the duration of the experiment. It can easily be shown that within sensible limits of the model's parameters, the effect of truncation will be to provide an estimate of fishing mortality which is too high. The amount of bias introduced will depend on the actual values of F, M, N and T. Another problem, but one less easy to examine, is the effect of truncation on the estimated variances and covariance of F and M.

The authors are currently examining the effects of truncation and, as an intermediate solution, a small simulation study is reported here. It is hoped that this study will first assess the adequacy of the approximate formulae (7a–c) and secondly demonstrate the effect on the estimates of terminating the recording of tag recoveries.

Samples of tag returns in successive time periods of unit duration were generated by using the fact that the (constant) probability of a tagged fish being caught in an interval is

$$P = F[1-\exp(-Z)].$$

The actual number recaptured in the ith time interval is binomially distributed with parameters P and N_i where N_i is the number of fish alive at the beginning of the interval. This binomial distribution was approximated by a Normal distribution to calculate P and its equivalent probability for natural mortality with the same mean and variance. The values of F and M used in the simulation were the values of $\hat{F}_c$ and $\hat{M}_c$ obtained using the data given in Table 1, namely $\hat{F}_c = 0{\cdot}38$ and $\hat{M}_c = 3{\cdot}23$. Two series of recoveries were generated, each consisting of 1000 sets of returns. In one set $T = 21$ (months) and in the other $T = 12$. Empirical distributions of $\hat{F}_c$ and $\hat{M}_c$ were derived in this way

TABLE 2. Estimates of F and M together with their variances and covariance calculated (a) from the data in Table 1 using equations (6a–b) and (7a–c); (b) from the empirical distribution generated by simulation letting the experiment run for $T = 21$ months; (c) as (b) but letting the experiment run for $T = 12$ months

Estimate	From Table 1	Simulation with $T = 21$	Simulation with $T = 12$
$\hat{F}_c$	0·38	0·38	0·41
$\hat{M}_c$	3·23	3·24	3·56
Var $(\hat{F}_c)$	0·0004	0·0004	0·0004
Var $(\hat{M}_c)$	0·0152	0·0142	0·0156
Cov $(\hat{F}_c, \hat{M}_c)$	0·0016	0·0016	0·0017

from which means, variances and covariances were calculated. These are given in Table 2 together with the corresponding values obtained from the actual recovery data given in Table 1. From these quantities 95% confidence ellipses for $\hat{F}$ and $\hat{M}$ have been derived and are shown in Fig. 1. As may be seen $\hat{F}_c$ and $\hat{M}_c$ are virtually unbiased when $T = 21$. However, as anticipated, when $T = 12$ both $\hat{F}_c$ and $\hat{M}_c$ are biased upwards. The simulation also

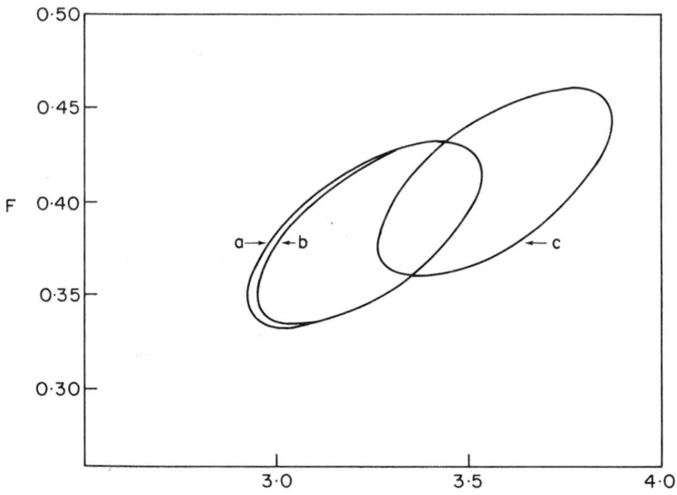

FIG. 1. The 95% confidence ellipses for F and M: (a) from the data given in Table 1; (b) from the simulation with $T = 21$; (c) from the simulation with $T = 12$.

supported the adequacy of the formulae (7a–c), at least for N large, which can be seen by the similarity of the values of the variances and covariances given in Table 2, and graphically by the almost identical confidence ellipses. The ellipse for $T = 12$ has simply been displaced by the bias in $\hat{F}_c$ and $\hat{M}_c$. An examination of the actual distributions of the individual values of $\hat{F}_c$ and $\hat{M}_c$ generated by the simulations showed them to be bivariate Normal and that the 95% confidence ellipses did in fact contain 95% of the observations.

3. Diffusion Model

The previous section considered the reliability of estimates derived from the fixed-parameter model and the robustness of these estimates when the underlying assumptions have not been met. This section describes a model for which the assumptions are less strict in that the fishing mortality may vary. Emphasis is again placed on providing a means of assessing the accuracy of derived estimates.

The essential features of the model are that the tagged fish disperse away from the point at which they are liberated, and are recovered in the catches of commercial fishing vessels. The rate at which they are recovered will depend on where they disperse to, and on the intensity of fishing effort which they encounter there. Fishing mortality is no longer constant, but varies with fishing effort in time and space.

The model is essentially that described by Skellam (1951), and the authors are indebted to Mr. Hans Lassen of the Danish Institute for Fishery and Marine Research for his formulation of the model in the context of fish population dynamics.

The fishing area is represented as a grid of equal-sized squares between which the fish may pass. The boundaries of the area may be either absorbing or reflecting. The probability p_{ij} of a fish in square j moving to square i is generated from a bivariate Normal distribution, i.e.

$$p_{ij} = \iint_A \frac{1}{2\pi\sigma^2} \exp\left\{\frac{-1}{2\sigma^2}[(x-\mu_x)^2+(y-\mu_y)^2]\right\} \mathrm{d}x\,\mathrm{d}y,$$

where the area of integration A is determined by the position of square i relative to square j. The parameters, σ, μ_x and μ_y determine the rate and direction of movement of the overall population.

Fishing mortality varies with time and space and is determined by

$$F_{ik} = qE_{ik}$$

where q is the "catchability" coefficient and E_{ik} the fishing effort in area i at time k. The probabilities of being recaptured and of natural mortality, including other losses, are then given by

$$\xi_{ik} = qE_{ik}\{1-\exp[-(M+qE_{ik})\,t]\}/(M+qE_{ik}),$$
$$\eta_{ik} = M\{1-\exp[-(M+qE_{ik})\,t]\}/(M+qE_{ik}),$$

respectively.

Within each time period, fish are first subjected to the two sources of mortality, and the survivors then spatially redistributed. If $N^{(k)}$ is the distribution over area of fish alive at the beginning of time period k and $n^{(k)}$ the distribution of recaptures, we can write

$$n^{(k)} = \boldsymbol{\xi}^{(k)} N^{(k)}$$

where $\boldsymbol{\xi}^{(k)}$ is a matrix with elements ξ_{ik} along the diagonal and all off-diagonal elements zero. Writing $\boldsymbol{\eta}^{(k)}$ to be the equivalent matrix for natural mortality, we have

$$N^{(k+1)} = \mathbf{P}(\mathbf{I}-\boldsymbol{\xi}^{(k)}-\boldsymbol{\eta}^{(k)})\,N^{(k)},$$

where $\mathbf{P} = [p_{ij}]$ is the spatial transition matrix for redistributing fish over area.

The problem now is to choose some criterion by which the model's parameters may be estimated. It is also essential that this criterion should provide some assessment of the validity of these estimates.

The tags used in this experiment were internal and were not detected at capture. The predicted tags returns must therefore be summed over area to be comparable with the observed distribution of tag returns.

Writing n_k as the predicted total number of tag returns in the kth time interval and r_k as the observed number of tags, then one method of estimation would be to minimize the sum over k of squared deviations between n_k and r_k. Referring to Table 1, however, it can be seen that the distribution of tag returns is very irregular. In fact more than 70% of the tag returns are accounted for in 2 of the 21 months. Since the largest absolute deviations tend to occur with the largest number of returns, the overall goodness of fit tends to be dominated by the fit in these 2 months. To avoid this, the expression

$$S = \sum_k (n_k - r_k)^2/r_k$$

was chosen to be minimized. This will be recognized as the minimum χ^2 criterion, which is effectively minimizing the sum over time of the squared deviations per tag returned.

This criterion can also be used to assess the validity of the derived estimates. Suppose we restrict our attention to the estimation of F and M. The values which minimize S for the data in Table 1 were $F = 0{\cdot}36$ and

$M = 0{\cdot}90$, where F has been calculated using the average fishing effort in the area over the time of the experiment. To correspond with the 95% confidence ellipses which were obtained for the fixed-parameter model, we define a 5% "acceptance region" for these estimates of F and M to be that region within which the increase in the minimum value attained by S is not greater than 5%. This region cannot be interpreted in any probabilistic sense, but demonstrates the range of values of F and M for which the increase in the minimum value of S would be below some acceptably small amount. This region also turns out to be an ellipse and is shown in Fig. 2.

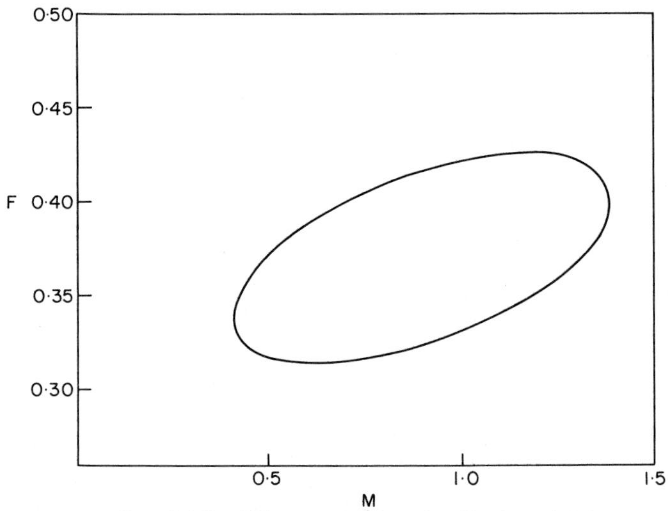

FIG. 2. The 5% acceptance region for F and M.

4. Summary

Two models have been presented, each of which provides estimates of both fishing and natural mortality rates for an exploited fish stock. The properties of estimates derived from these models have been illustrated with data collected from a tagging experiment carried out on herring in the south-eastern North Sea.

Attention has been focused primarily on the problems of making inferential statements about the true mortality rates and, for one of the models, the robustness of these statements when the assumptions underlying the model are not met.

REFERENCES

CHAPMAN, D. G. (1961). Statistical problems in dynamics of exploited fisheries populations. *Proc. 4th Berkeley Symp.* 1960, **4**, 153–168.

CORMACK, R. M. (1968). The statistics of capture–recapture methods. *Oceanogr. Mar. Biol. Ann. Rev.* **6**, 455–506.

GULLAND, J. A. (1955). On the estimation of population parameters from marked numbers. *Biometrika* **42**, 269–270.

ICES (1975). Report on the ICES Working Group on the Bløden Tagging Experiment, 1969/70. *Coop. Res. Rep. Int. Coun. Explor. Sea* **47**, 137.

JONES, R. (1966). Manual of methods for fish stock assessment. Part IV. Marking. *FAO Fish. Tech. Paper No.* 51, *Suppl.* 1. *FRS/T*51 *Suppl.* 1 *MFS Marking*.

PAULIK, G. J. (1963). Estimates of mortality rates from tag recoveries. *Biometrics* **19**, 28–57.

POPE, J. A. (1963). The design and analysis of capture–recapture experiments. International Commission for the Northwest Atlantic Fisheries. *Special Publication* No. 4, North Atlantic Fish Marking Symposium Woods Hole, Mass., May 1961, pp. 342–347.

SKELLAM, J. G. (1951). Random dispersal in theoretical populations. *Biometrika* **38**, 196–218.

6. The Partitioning of Natural Mortality in a Multispecies Model of Fishing and Predation

K. P. ANDERSEN AND ERIK URSIN

Danish Institute for Fishery and Marine Research, Charlottenlund, Denmark

1. Introduction

In the "Simple Theory of Fishing" (Beverton and Holt, 1957) the mortality of the individual fish species considered is introduced by

$$\frac{dN}{dt} = -ZN, \quad Z = F + M,$$

where N is the number, t is time, F is the instantaneous coefficient of fishing mortality and M similarly the coefficient of "natural" mortality, i.e. all other mortality.

M soon proved an important parameter difficult to estimate with acceptable accuracy. Little has been done, unfortunately, to disclose the biological phenomena behind the parameter because, in a one-species model, it is often of little consequence how the mortality is brought about if only its size is known.

Work on multispecies models involves interactions between fish species and their relations with invertebrates. This requires definite assumptions on the sources of mortality and the way they operate. We shall discuss some problems encountered at the Danish Institute for Fishery and Marine Research during the development of a multispecies extension to the Beverton and Holt model. (A list of unpublished documents is given at the end of the paper. These are available, on request, from the authors.) The model maintains a mass balance necessitating a predefined destination of animals disappearing in other ways than by fishing depending on whether they emigrated, were eaten or just died. An outline of the model is given in the Appendix to this paper. Obvious sources of mortality (including migration) are the following.

1. *Animals removed from the system*
   (a) Fishing proper (fish landed).
   (b) Emigration interpreted as mortality.

2. *Animals added to the system*
   (a) Immigration interpreted as negative mortality.

3. *Animals remaining in the system after death arising from the following*
   (a) Predation (nutrient contents of prey partly recirculated, partly incorporated in predator).
   (b) Starvation (scarcely relevant in adults).
   (c) Diseases including parasites (which might also be included in the predators).
   (d) Old age.
   (e) Spawning strain.
   (f) Adverse physical conditions (e.g. cold winters).
   (g) Poisons.
   (h) Discard at sea (animals caught but discarded and dead from injuries suffered).
   (i) Escape in injured condition (fatigue, guts crushed at passage during meshes, etc.).

All of these mortalities except fishing mortality proper are included in most estimates of M, although there has recently been a tendency to include discard mortality (but not escape mortality) in F.

In the multispecies model we put

$$M = M1 + M2 + M3 + \ldots,$$

each term accounting independently for some source of mortality or apparent mortality. Some terms retain the parameter character of the Beverton and Holt M whereas others are computed dynamically in the model and are therefore not parameters themselves. Only the latter are of interest in the present context. They are

Mortality due to predation;
Larval mortality maintaining the stock-recruitment relationships;
Mortality due to poisonous pollutants;
Apparent mortality due to migration.

2. Mortality due to Predation

To describe predation (or grazing) mortality of one species is to describe the food consumption of other species. In the outline of the model (Appendix) this is achieved by the expression

$$M2_i = \left(\sum_j \frac{\mathrm{d}R_j}{\mathrm{d}t} N_j \frac{\phi_{ji}}{\phi_j} \Big/ N_i w_i \right), \quad i = 1, 2, \ldots, n, \quad j = 1, 2, \ldots, n.$$

This is the instantaneous mortality rate of i being preyed upon by j's. n is the number of categories (species, age groups, etc.) accounted for in the model. The denominator is the total biomass of the prey i and the numerator the biomass of i consumed in time $\mathrm{d}t$. $\sum(\mathrm{d}R_j/\mathrm{d}t)\,N_j$ is the total consumption by all species in the system. ϕ_j is the food available to j, ϕ_{ji} the part of this food which is i. At first sight it may seem as if j eats all kinds of prey indiscriminately, but this is not so because the available food for j (i.e. $\phi_j = \sum_i \phi_{ji}$) is already weighted by the suitability index G_{ji} of i as food for j (see Appendix).

The suitability index must be considered in more detail. The *food of i*, G_{ij}, can be factorized into an index g_{ij} of the suitability of an animal of j's size as food for an animal of i's size, and an index of vulnerability ρ_{ij} of j to i (e.g. j may be inaccessible to i because buried in sand). Thus:

$$G_{ij} = \rho_{ij}\, g_{ij}.$$

Consider first the size ratio problem. Investigations on cod and dab (Ursin, 1973) and copepods (Ursin, 1974a) indicate that a predator has a preferred prey size and that a prey twice as big as the preferred size is as attractive as another prey half the preferred size. This suggested a size-preference curve related to the log-normal probability function. The exponential part is sufficient when prey of the preferred size is allotted a value of 1:

$$g_{ij} = \exp\left(-\frac{\ln(w_i/w_j)-\eta_i{}^2}{2b_i^2}\right), \quad 0<g_{ij}\leqslant 1.$$

η_i is the logarithm of i's "preferred" ratio w_i/w_j; b_i^2 is a parameter indicating how particular i is in its selection of prey. Jones (in press) found that young cod take prey 8·3 times the size of the prey of haddock and 6·8 times the size of that of saithe. With $\eta_{\mathrm{cod}} = 5{\cdot}1$ (Ursin, 1973) we get $\eta_{\mathrm{haddock}} = 7{\cdot}2$ and $\eta_{\mathrm{saithe}} = 7{\cdot}0$.

The curve described by g_{ij} may be adjusted at the ends of its range in accordance with the predator's anatomy when there is an upper or a lower limit to physically feasible prey size.

Turning to the vulnerability index, consider a matrix ρ of vulnerability indices ρ_{ij} of j as prey for i. Assume the values adjusted so that $\max_j\{\rho_{ij}\} = 1$. Then with $\rho_{ij} = 1$ and $w_j = w_i/\exp(\eta_i)$, j contributes to the available food ϕ_i of i with its biomass: $\phi_{ij} = N_j w_j$. In all other cases it contributes less.

According to Jones (in press), cod is well adapted to taking moving prey whereas stationary prey is more vulnerable to haddock. This would appear in the row vectors $\bar{\rho}_i$ of ρ for cod and haddock (although the elements ρ_{ij}

are not estimated by Jones). The mechanism, although perhaps slightly more rigid, resembles that visualized by Jones. The details are not yet analysed in computer runs because we hesitate to introduce such a large array of elements ρ_{ij}, most of which would have to be guessed. Instead we use a column vector $\bar{\rho}_j$ indicating the vulnerability of each category to predation irrespective of the predator.

Related to the vulnerability pattern is the pelagic/demersal behaviour of animals. It is conveniently expressed by the fraction p_i of the population of i which is pelagic at any moment. Species (or developmental stages of a species) may be either purely pelagic ($p_i \equiv 1$), purely demersal ($p \equiv 0$) or optional ($0 \leqslant p \leqslant 1$), distributing themselves according to the availability of food in the pelagic and the demersal habitats. In the program the calculation is based on the quantities of purely pelagic and purely demersal food. There are then two biomasses of food available to i: $\phi_i{}^{\mathrm{P}}$ (pelagic food) and $\phi_i{}^{\mathrm{D}}$ (demersal food):

$$\phi_i{}^{\mathrm{P}} = \sum_j p_j G_{ij} w_j N_j, \quad \phi_i{}^{\mathrm{D}} = \sum_j (1-p_j) G_{ij} w_j N_j.$$

Individual terms may be zero.

The feeding level (Appendix) must be calculated separately for the two habitats:

$$f_i^{\mathrm{P}} = \frac{\phi_i{}^{\mathrm{P}}/V}{(\phi_i{}^{\mathrm{P}}/V)+1/q_i{}^{\mathrm{P}}}, \quad f_i^{\mathrm{D}} = \frac{\phi_i{}^{\mathrm{D}}/A}{(\phi_i{}^{\mathrm{D}}/A)+1/q_i{}^{\mathrm{D}}},$$

where V is water volume and A the area of the sea floor. The search rate parameters $q_i{}^{\mathrm{P}}$ and $q_i{}^{\mathrm{D}}$ are different because search is by volume per unit time in the pelagic habitat and by area per time unit in the demersal habitat. f_i^{P} or f_i^{D} may be zero. The average feeding level of i is a function of p_i:

$$f_i = f_i^{\mathrm{P}} p_i + f_i^{\mathrm{D}}(1-p_i).$$

Similarly, the predation mortality $M2_i$ depends on the distribution of the population of i between the two habitats and on how its predators j are distributed. By summing as before, but over rows instead of columns, we get:

$$M2_i{}^{\mathrm{P}} = \frac{1}{p_i N_i w_i} \sum_j \left(\frac{\mathrm{d}R_j{}^{\mathrm{P}}}{\mathrm{d}t} p_j N_j \frac{\phi_{ji}{}^{\mathrm{P}}}{\phi_j{}^{\mathrm{P}}} \right),$$

$$M2_i{}^{\mathrm{D}} = \frac{1}{(1-p_i) N_i w_i} \sum_j \left(\frac{\mathrm{d}R_j{}^{\mathrm{D}}}{\mathrm{d}t} (1-p_j) N_j \frac{\phi_{ji}{}^{\mathrm{D}}}{\phi_j{}^{\mathrm{D}}} \right),$$

where the rations consumed are calculated using f_i^{P} and f_i^{D}:

$$\frac{\mathrm{d}R_j{}^{\mathrm{P}}}{\mathrm{d}t} = f_j^{\mathrm{P}} h_j w_j{}^m, \quad \frac{\mathrm{d}R_j{}^{\mathrm{D}}}{\mathrm{d}t} = f_j^{\mathrm{D}} h_j w_j{}^m.$$

$M2_i$ becomes the weighted mean of $M2_i^{\mathrm{P}}$ and $M2_i^{\mathrm{D}}$:

$$M2_i = M2_i^{\mathrm{P}}\, p_i + M2_i^{\mathrm{D}}(1-p_i).$$

$M2$ values are generally high for small animals and negligible for the biggest animals because there are few other animals which can eat them.

3. Larval Mortality maintaining the Stock–Recruitment Relationship

The grazing mechanism described above proved insufficient to prevent the system from exploding in one way or another except when extreme cannibalism was introduced by making all kinds of animals much more vulnerable (10–100 times) to predation by their own species than by any other species. Although perhaps realistic in special cases this seemed unsatisfactory as a general description of the mechanism of a marine ecosystem.

Also it seemed necessary to indicate an upper limit of the stock size of each species because the physical environment is not yet elaborated enough in the model. There is, for example, no oxygen budget so that a "safety valve" is necessary to prevent biomass from increasing to such magnitudes that oxygen demand becomes critical (as would occur with too thick a layer of herring eggs on the sea floor).

Both problems found a (temporary) solution by introducing a mortality of the following type for each species of animal:

$$M1_i = \frac{N_i}{R_i}\,\frac{(\mathrm{d}w_i/\mathrm{d}t)_{\max}}{w_i},$$

where $(\mathrm{d}w_i/\mathrm{d}t)_{\max}$ is the highest possible growth rate at age t, and R_i is a recruitment parameter (a number of fish) maintaining the stock size. Disregarding for the moment the second part of the expression and other kinds of mortality, R is approximately the number of recruits surviving until 1 year of age because the number at 1 year (N_{i1}) is much smaller than the number of eggs (N_{i0}):

$$\frac{\mathrm{d}N_i}{\mathrm{d}t} = -M1_i\, N_i = -\frac{1}{R_i}N_i^2, \quad R_i = N_{i1}\frac{N_{i0}}{N_{i0}-N_{i1}}.$$

When N_i is high, mortality is high. When N_i is low so is mortality. This makes mortality a function of the number of eggs hatched. Using the von Bertalanffy growth equation, the second part of the expression (the maximal relative growth rate at age t) is higher the smaller the fish is. When food is scarce the growth is slow, the size remains small for a long time and the mortality remains high. This simulates a mortality due to starvation (an idea by Jones but not in his formulation). The starvation mortality increases when there are many competitiors of one's own species because N_i enters as a factor in the first part of the expression. This emphasizes another observation by Jones (in press) who found that differences in behaviour cause each species to compose a different diet so that interspecies competition is partly replaced by coexistence.

R_i should be capable of estimation outside the model although this has not yet been attempted. If we neglect the index i and put

$$\frac{\Delta N}{\Delta t} = -[M1(t) + \hat{M}(t)]\, N(t) = -\frac{1}{R} N(t)^2 \frac{(\Delta w/\Delta t)_{max}}{w(t)} - \hat{M}(t)\, N(t)$$

we can:

(1) Assume "other mortality" $\hat{M}(t)$ negligible using data for young larvae only because these would have a high $M1$ as they are very numerous. A plot of $\Delta N/\Delta t$ against $N(t)^2(\Delta w/\Delta t)_{max}/w(t)$ is obvious. As an alternative the corresponding differential equation could be solved:

$$\frac{1}{N(t)} = \frac{1}{R}\int_{t_0}^{t} \frac{(dw/d\tau)_{max}}{w(\tau)}\, d\tau + \frac{1}{N(t_0)}$$

and $1/N(t)$ plotted against the integral evaluated numerically.

(2) Assume $\hat{M}(t) = \hat{M}$ in the period of observation and apply multiple linear regression analysis to the difference equation with $1/R$ and $\hat{M}$ as the coefficients to be estimated.

(3) Assume $\hat{M}(t)$ otherwise estimated and add $\hat{M}(t)\, N(t)$ to the left side of the difference equation which could then be treated as in (1).

In any case the observations required are a time series of $w(t)$ and $N(t)$ as well as estimates of the growth parameters which are necessary for the evaluation of $(\Delta w/\Delta t)_{max}$.

A different approach to the problem of density-dependent larval mortality due to starvation is suggested by Beyer and Lassen (1975) in a time-discrete variety of the model. In the original model each age group of species i is represented by one number N_i and one body weight w_i, whereas Beyer and Lassen distribute N_i on a number of weight classes w_{ir} each greater than the preceding one by a constant factor b ($\sqrt{10}$ was used). Let Δw_{ir} be the average weight increment in time Δt of an animal with initial weight w_{ir}, and ΔK_{ir} its fasting metabolism measured as the weight loss of a fasting animal. If $\Delta w_{ir} = 0$, a simple procedure is to let all animals N_{ir} remain in weight class r. If $\Delta w_{ir} < 0$ and if r is the smallest weight class then the fraction

$$-\frac{\Delta w_{ir}}{\Delta K_{ir}}$$

of N_{ir} can be assumed to starve to death while the remaining members of N_{ir} maintain their weight and remain in the smallest weight class. The requirement for survival is easily strengthened, for example by requiring that an animal shall die if it does not increase its weight by a factor b in the time interval Δt. In that case the fraction dying from insufficient feeding is

$$-\frac{\Delta w_{ir}+w_{ir}(b-1)}{\Delta K_{ir}+w_{ir}(b-1)}.$$

The reasoning behind the expression is as follows.

Omitting for the moment the subscripts *ir*, the biomass of survivors under the constraint imposed by the factor b, is

$$N_1 bw = N_1(w+\Delta A-\Delta K),$$

where ΔA is food assimilated and ΔK the weight loss due to indispensable metabolism (the metabolism of a fasting animal). It follows that

$$\Delta A = \Delta K+w(b-1).$$

If there were no mortality the biomass would be

$$N_0(w+\Delta w) = N_0(w+\Delta\hat{A}-\Delta K)$$

from which $\Delta\hat{A} = \Delta w+\Delta K$ where $\Delta\hat{A}<\Delta A$ because the food is imagined distributed over a larger number of animals. Assuming that the survivors get all the food, we can put

$$N_1\Delta A = N_0\Delta\hat{A}$$

and find the fraction dying as

$$1-\frac{N_1}{N_0} = 1-\frac{\Delta\hat{A}}{\Delta A} = 1-\frac{\Delta w+\Delta K}{\Delta K+w(b-1)} = \frac{-\Delta w+w(b-1)}{\Delta K+w(b-1)}.$$

Survivors are moved up one class because b is also the size ratio of successive weight classes. This approach seems a realistic way of introducing larval mortality as described by Jones (1973) into a deterministic model. Unfortunately, the number of groups to be handled increases considerably by introducing several weight classes for each age group.

4. Mortality due to Poisonous Pollutants

The animals in the time-continuous model are able to absorb and eliminate a persistent (non-degradable) chemical compound such as DDT or PCB. Absorption is from the water or from the food. Elimination occurs either as metabolites or as the original compound. The compound follows the biomass in general, not any specific tissue. Its pathways in the ecosystem are analogous to those of the food of animals (Lassen, in press; Beyer and Lassen, 1975).

Mortality due to the pollutant is assumed proportional to the concentration of pollutant in the animal. With p_i the weight of the pollutant in an animal of group i weighing w_i, we have

$$M3_i = \widehat{M3}_i \frac{P_i}{w_i}.$$

More important, perhaps, than mortality due to poisoning is an increased vulnerability of the affected animals to disease and predation. This speeds up the rate of transfer of pollutants through the food chain. The assumptions involved are not yet formulated, but the problem is being worked on (H. Lassen, personal communication).

5. Migration

In many cases migration can be treated as a mortality with positive or negative sign corresponding to the transport coefficient T of Beverton and Holt (1957).

Let the sea in question consist of two homogeneous boxes and consider the effect of migration upon the population of i in box 1. Let $T_{1,i}$ be the coefficient of transport from box 1 to box 2 and $T_{2,i}$ from box 2 to box 1. The populations of i are represented by $N_{1,i}$ and $w_{1,i}$ in box 1 and by $N_{2,i}$ and $w_{2,i}$ in box 2. Then

$$\frac{dN_{1,i}}{dt} = \left(\frac{dN_{1,i}}{dt}\right)_u - T_{1,i} N_{1,i} + T_{2,i} N_{2,i},$$

$$\frac{dw_{1,i}}{dt} = \left(\frac{dw_{1,i}}{dt}\right)_u + T_{2,i} N_{2,i}(w_{2,i} - w_{1,i}) \Big/ N_{1,i},$$

where the subscript u indicates the usual rates of change as there would be if transport between boxes did not occur. If the specimens of i are the same size in both boxes the last term of the weight equation vanishes. If they are different sizes the biomass difference is distributed evenly over the $N_{1,i}$ specimens in box 1. The grazing pattern in box 1 changes because who-eats-whom is partially determined by body size ratios. With Beyer's amendment to the model (several weight classes in each i, as described above) immigrants from box 2 join their own weight class in box 1 and influence the grazing pattern in their own right only; the size of the animals already in box 1 remains unchanged.

Appendix. Outline of a Multispecies Extension to the Beverton and Holt Model

(as described by Andersen, Lassen and Ursin, 1973)

The Beverton and Holt (1957) simple theory of fishing has three basic differential equations in time for each species i:

$$\frac{dw_i}{dt} = H_i w_i^m - k_i w_i^n, \tag{1}$$

$$\frac{dN_i}{dt} = -(F_i + M_i) N_i, \tag{2}$$

$$\frac{dY_i}{dt} = F_i N_i w_i; \tag{3}$$

(1) is the change in body weight as a function of body weight and the rates of food intake and metabolism;

(2) is the change in numbers as a function of numbers and of fishing and natural mortality;

(3) is the yield of the fishery. F is proportional to the fishing intensity.

Include a fourth equation:

$$\frac{dR_i}{dt} = f_i h_i w_i^m, \quad 0 \leqslant f_i \leqslant 1, \tag{4}$$

where R_i is food consumed by the species in question. If we put the rate of food consumption proportional to w^m this agrees with the positive term of (1). The proportionality constant is h_i when the fish eats all it can. When it is not allowed to do so it eats the fraction f_i of the maximum. f is the feeding level ($\equiv \xi/\xi_L$ in the notation of Beverton and Holt, 1957). The fraction v of food consumed ends up in the fish body, i.e. it equals the positive term of (1).

Thus,

$$\frac{dw_i}{dt} = v_i \frac{dR_i}{dt} - k_i w_i^n = v_i f_i h_i w_i^m - k w_i^n, \tag{1A}$$

so that $H = vfh$.

The feeding level is determined by the expression

$$f_i = \frac{\phi_i/V}{\phi_i/V + (1/q_i)}, \tag{5}$$

where V is the water volume. $1/q_i$ is known in the literature on enzyme kinetics as the half-saturation constant. q_i is a coefficient of the rate of search for food. ϕ_i is the sum of biomass of all kinds of prey, each weighted by its suitability G to the predator:

$$\phi_i = \sum_j \phi_{ij} = \sum_j G_{ij} w_j N_j. \tag{6}$$

The suffix j designates the species or group consumed by the predator i. That deals with the necessary amendments to the growth sector of the Beverton and Holt model.

The mortality sector as described by (2) also needs refinement. The natural mortality coefficient M of Beverton and Holt covers all sources of mortality except fishing mortality. To describe interspecific relationships, M must be partitioned into $M = M1 + M2$ where $M2$ is the grazing, or predation, mortality:

$$\frac{dN_i}{dt} = -(F_i + M1_i + M2_i)\,N_i, \qquad (2A)$$

where $M2_i$ is the fraction consumed of the biomass of i by the biomass of all groups preying upon i. With the terminology introduced above this can be formalized as

$$M2_i = \frac{1}{N_i w_i} \sum_j \frac{dR_j}{dt} N_j \frac{\phi_{ji}}{\phi_j}. \qquad (7)$$

The main achievement of the Beverton and Holt model as the basis of a multispecies stock and yield model is the precise account of—and the dynamic relations between—food consumed and predation mortality expressed by equating the instantaneous rate sum for all species in the system:

$$-\sum_i \left(\frac{dN_i}{dt}\right)_{\text{predat}} w_i = \sum_i \frac{dR_i}{dt} N_i$$

or

$$\sum_i M2_i w_i N_i = \sum_i f_i h_i w_i^m N_i.$$

This is particularly useful when a complete account of phosphorus circulation is kept.

The model is used in the analysis of North Sea fisheries management problems caused by recent changes in fish stock composition which may be caused, for example, by poisonous pollutants or eutrophication, or (and perhaps more likely) by the fishery itself because the reduction of certain stocks by the fishery changes predation patterns and food concentrations. Various computer programs (Andersen and Ursin, in press; Ursin and Andersen, in press; Lassen, in press) have been worked with from 14 to 18 categories (subscripts i or j) of plants, animals and nutritive matter, calling for the simultaneous solution of 42–274 differential equations.

References

ANDERSEN, K. P. and URSIN, E. A multispecies analysis of the effects of variations of effort upon stock composition of eleven North Sea fish species. ICES Symposium (1975) on the Changes in the North Sea Fish Stocks and their Causes. *Rapp. P.-v. Réun. Cons. Int. Explor. Mer* (in press).

BEVERTON, R J. H. and HOLT, S. J. (1957). On the dynamics of exploited fish populations. *Min. Agric. Fish. Food. Fishery Inv.* Ser. 2, Vol. 19, 531pp. London.

JONES, R. (1973). Stock and recruitment with special reference to cod and haddock. *Rapp. P.-v. Réun. Con. Int. Explor. Mer*, **164**, 156–173.

JONES, R. Some observations on competition between some gadoid fish species. ICES Symposium (1975) on the Changes in the North Sea Fish Stocks and their Causes, *Rapp. P.-v. Réun. Cons. Int. Explor. Mer* (in press).

LASSEN, H. A method of predicting the biological effects of synthetic pollutants in the North Sea. ICES Symposium (1975) on the Changes in the North Sea Fish Stocks and their Causes, *Rapp. P.-v. Réun Cons. Int. Explor. Mer* (in press).

URSIN, E. (1973). On the prey size preferences of cod and dab. *Medd. Danm. Fiskeri-og Havunders.*, NS, **7**, 85–98.

URSIN, E. and ANDERSEN, K. P. A model of the biological effects of eutrophication in the North Sea. ICES Symposium (1975) on the Changes in the North Sea Fish Stocks and their Causes, *Rapp. P.-v. Réun. Cons. Int. Explor. Mer* (in press).

OTHER REFERENCES

Mimeographed papers dealing with various aspects of the multispecies model.

ANDERSEN, K. P., LASSEN, H. and URSIN, E. (1973). A multispecies extension to the Beverton and Holt assessment model with an account of primary production. ICES C.M. 1973/H:20.

ANDERSEN, K. P. and URSIN, E (1975). Interspecies competition in fish. Read at a seminar in Lowestoft, 28 November 1975. Privately distributed.

BEYER, J. E. and LASSEN, H. (1975). Multispecies models of an exploited sea. Symposium on Arctic Systems, St. John's, 18–22 August 1975, 64pp.

LASSEN, H. and NIELSEN, P. B. (1972). Simple mathematical model for the primary production as a function of the phosphate concentration and incoming solar energy applied to the North Sea. ICES C.M. 1972/L:6, 21pp.

URSIN, E. (1974a). Search rate and food size preference in two copepods. ICES C.M. 1974/L:23, 13pp.

URSIN, E. (1974b). Input data for a multispecies yield assessment model of North Sea fisheries. ICES C.M. 1974/H:40, 14pp.

7. Analytic Approaches in Fishing Technology

J. J. Foster, R. S. T. Ferro and A. J. Reid

Marine Laboratory, Aberdeen

1. Introduction

This paper describes some analytical methods for solving fishing technology problems. Emphasis is given to fishing gear in the context of developing a mathematical model for the complete fishing system.

Fisheries models are needed to aid the rational deployment of the extensive capital invested in fishing vessels and processing plant. Some simple overall models have been produced (F.A.O. Report, 1975) in which the fishing system is divided into blocks. The fishing gear block and its relationship to other blocks are considered here. In particular prediction techniques developed in trawl gear research have been applied in formulating the gear model and a full practical example is cited.

Relevant techniques for gear performance measurement and observation are considered and shown to contribute to the refinement of the model.

Whilst the fishing industry provides this special and interesting background in which to carry out technological research, progress with modelling is slow because a significant proportion of the available effort has to be applied to immediate problems. Answers are required quickly and until the management tool of a comprehensive model is prepared we are forced to use the submodels and other techniques described.

2. The Fishing System

In order to quantify the economic viability of a proposed fishery a number of factors must be evaluated. Broadly these factors are:

(a) existence of fish stocks including detailed information on species, density, seasonal changes, behaviour patterns, etc.;

(b) effectiveness of catching operations including gear(s) and vessels together with knowledge of distances from grounds to landing port(s);

(c) landing facilities;

(d) processing facilities;

(e) distribution facilities;
(f) markets.

Most of these are highly dependent on the availability of manpower and investment capital.

Traditionally, attempts to model the fisheries have concentrated mainly on one of these factors. The treatment of this main factor has often been considerably refined whereas treatment of the others has been very simple and idealistic. Most models used vessel size, hold capacity and/or engine power as the main variables. Undoubtedly this occurred because vessel technology had reached a relatively advanced stage, the vessel parameters could be readily defined, and the vessels were an important part of the initial investment and running cost.

Although simple models of this sort have been available for some time (F.A.O. Report, 1975) they are little used, administrators making intuitive decisions based on isolated facts, stock assessments and market demand forecasts.

Where the models are used, serious limitations are evident as in the case given below of matching gear design to vessel power for a particular fishery (see also Engvall and Engstrom, 1969). In this case simplification is partly responsible for the difficulties. The arguments can be summarized as follows. The basic formula used to relate vessel and gear parameters is

$$\text{S.H.P.} = CAT_d V^2 \tag{1}$$

where S.H.P. = shaft horse power, T_d = mean twine diameter over the whole net, A = mouth area of net, calculated as net spread × net height, V = trawling speed and C is a constant to take care of trawl design and propulsive efficiency.

By an analysis of current fishing activity an empirical relationship between T_d and S.H.P. is derived for existing vessels and this relationship, substituted into equation (1), yields the following equation.

$$A = \frac{1}{CV^2}\left(\frac{1}{0{\cdot}00123} + \frac{1{\cdot}3}{\text{S.H.P.}}\right). \tag{2}$$

Then, from limited measurements of a specific trawl, a table of spot values is constructed and equation (2) used to evaluate C

A (m²)	S.H.P.	V (knots)	C
115	400	3	0·7855
146	575	3	0·6187
150	600	3	0·6022
		Mean value =	0·6688

The mean value of C, so calculated, is then substituted into equation (2).

$$A = \frac{1}{1 \cdot 495 V^2}\left(\frac{1}{1 \cdot 00123} + \frac{1 \cdot 3}{\text{S.H.P.}}\right). \tag{3}$$

Equation (3) is used to relate net mouth area to vessel speed and shaft horse power and, in this form, is applied with very doubtful justification to a range of vessels and over a range of trawling speeds.

The next step is to set catch rate to be directly proportional to the volume of water sieved per unit time, using equation (3) to calculate the sieved volume.

It is clear that such an arbitrary method of deriving a relationship between gear and vessel and then catch and vessel is unlikely to be accurate. Not surprisingly, lack of success with such oversimplified approaches has forced reappraisal.

Many workers concerned with global models of a fishery are now agreed (F.A.O. Report, 1975) that each of the main factors listed above in the total fishing model is sufficiently complex that it should be treated separately and comprehensively.

3. The Fishing Gear Block

The main aim of this paper is to describe some of the work done in the field of fishing gear technology, the results of which may be used in the design of the appropriate block in a global fisheries model.

There are many types of fishing gears but we are concentrating our effort on trawling as opposed to other methods. A typical model of the fishing gear block can be divided into three parts, engineering, fish behaviour and economics.

(1) *Engineering*

This part involves a mathematical description of the engineering parameters describing the geometry of the gear and the internal and external forces, and will be described in detail later.

(2) *Fish behaviour*

While the principles apply to other methods the trawl is the easiest to explain. The objective is to determine the efficiency of a gear at catching the available fish.

This has been described mathematically (Foster, 1969) in an expression for catch in terms of the efficiency of each gear component.

Each part of the gear (Fig. 1) in turn approaches the fish which may be influenced towards capture or escape. The ratio of those herded towards capture (or captured in the case of the net) compared to the total approached is defined as the efficiency, $\mathscr{E}$, of that component.

In the horizontal plane

$$C = \rho A(\sum l_i \mathscr{E}_i)\, vt,$$

where C = catch, ρA = density of fish in fishing area, l_i = length of gear component i (calculated as the projected length perpendicular to the direction of motion), v = fishing gear speed, t = effective fishing time.

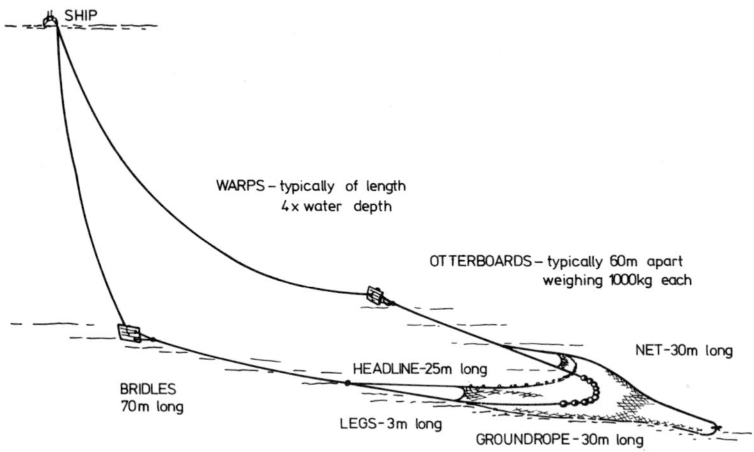

FIG. 1. The main components of a trawl gear.

Observation has shown that the angle of attack of a component influences the reaction of fish. The efficiency ($\mathscr{E}$) is therefore varied according to the angle of attack. At present poor estimates of $\mathscr{E}$ are used but behaviour studies involving sector-scanning sonar and fish-tracking techniques are already contributing more information to this subject. In addition, comparative fishing techniques using simple gears are reducing the range of uncertainty.

(3) *Economics*

For each gear the total cost of production and a maintenance cost must be estimated.

4. Output of Fishing Gear Model

The important output from the fishing gear block is data on the vessel power required to operate the fishing gear and the predicted catching rates of the various types of gear suitable for a given fishery. This information, when coupled with data on vessel performance, enables catching rate per vessel gear combination to be predicted and an estimate made of the fishing effort which will achieve the maximum sustainable yield. For each gear type the output is given in the form of a table of power requirements, operating costs and catch rates for a range of gear sizes.

5. Trawl Gear Models

The gear block has been discussed in relation to the general fishery problems. Detailed attention will now be given to engineering aspects of trawl gear models. The main components of a typical demersal trawl gear are shown (Fig. 1) and include the ship, warps, otter boards, bridle wire systems and the net.

The construction of a trawl gear model requires a knowledge of the forces acting on the gear and of the relationships between the forces and moments acting within the system, namely the equations of static or dynamic equilibrium. "Theoretically", a model based on such information will allow complete definition of the gear system given only the physical dimensions of each component and the prevailing environmental conditions.

The equations of equilibrium can be written in general terms without trouble. However, the laws governing the generation of the forces involved in these equations are not easily stated. For instance, insufficient is known about the forces on a vibrating cable towed through the water, the forces on gear components in contact with the sea bed, or the forces generated by flow through netting at angles of incidence of about 10–20° to the flow. At present it is necessary to rely on experimental data to provide information of a general nature which can then be applied to specific situations and which ideally will elucidate the general principles involved.

Such a model has been developed in collaboration with the British Hovercraft Corporation (Crewe, 1975) over a number of years. The following brief description of this type of approach is based on this work.

To obtain a model of the whole system it is necessary to build models of all the individual gear components which are subsequently interfaced with appropriate constraints at each junction. Special attention is given below to the net which typifies problems which occur with the other components of the gear.

Wire system models

The model assumes that five forces act on short, straight elements of the towed wire; tensions at either end, weight and two hydrodynamic forces in the plane of the local water velocity given in terms of skin friction and pressure coefficients. The coefficients may depend on the lay of the wire and its vibration. By starting at the forward end of the wire with a given tension vector, the hydrodynamic forces and weight acting on the first element and hence the change in the tension vector are calculated. The latter is input to the next element and so on.

In the general case, to obtain a solution for the complete wire configuration, empirical values for the two hydrodynamic coefficients and an initial magnitude and direction for the tension at the ship end of the gear are required. The latter tension is commonly not available. Thus an iteration technique is employed to converge on three suitable known quantities after choosing a very approximate initial value for the tension at the ship end. A Newton–Raphson convergence routine is used which in almost all cases reaches the solution in two or three iterations.

It is by no means sufficient to assume, as some workers have done in the past, that the shape of the various towed wires used in fishing gear are straight. For example, even at normal warp lengths of 300 fathoms, although the curvature may be small, the different assumptions have a very significant effect on the calculated lateral spread of the gear. However, in deep water on the continental slope the length of each warp may be up to 1000 fathoms and in this case the assumption that the warps are straight may give rise to totally misleading results.

Otter board models

Otter boards are sufficiently different from standard aerofoil shapes that model tests in wind tunnels or towing tanks have had to be used to determine the hydrodynamic forces. It is only recently, for instance, that data have been obtained on the effect of heeling and tilting on the coefficients of lift and drag. Representation of the interaction between an otter board and the sea bed is particularly difficult because of the uncertainty concerning sea bed composition. There is often no definite demarcation between the water and the sand or mud layer into which the otter board will sink to some extent. Some simplifying approximation can be made but the model may then be insufficiently representative considering the significant effect that sea bed type may have on gear performance. In the case of pelagic trawling where no part of the gear touches the sea bed, modelling of otter board behaviour is much simplified.

Net models

A considerable amount of research has been put into the modelling of nets, so far, without notable success. A totally theoretical approach is to consider netting as a specific example of a porous surface. Some research has been done into fluid flow across porous surfaces; in particular (Hamilton, 1970) has studied the flow into wedges and cones with porous walls. This approach is still a long way from understanding and describing the flow through a full-scale symmetrical fishing net with all its added complications of floats, weights and other appendages.

The important properties of a fishing net are its hydrodynamic drag and lift forces for these not only dictate the power of vessel required to tow the net but also determine the geometry. A more practical approach to the study of nets is to use actual pieces of netting and small-scale nets in towing tanks and wind tunnels, where the basic parameters drag, lift and geometry can be measured.

The use of water tank and wind tunnel to test fishing nets

In making net models and performing experiments upon them a major difficulty lies in choosing which criteria to use to make scale models. There are two techniques.

(i) Design the experiment to operate over a range of Reynolds numbers occurring in practice. This would maintain the same ratio of inertial to viscous forces for both model and full-size net. If tests are to be performed in water then this technique requires that the velocity of the model should be inversely proportional to the scale of the model. Unless very large models are constructed the water speeds required for such tests are usually impractical. Thus this technique is usually confined to tests on flat panels or simple shapes of netting (such as cones, cylinders, etc.). These tests may be done in wind tunnels where suitable fluid speeds can be achieved (Arlotte and Crewe, 1965; Fridman, 1973; Kowalski and Giannotti, 1974).

To apply this technique to the determination of net drag (and lift) an actual net is expressed as a combination of flat panels or simple shapes. The total net drag is then calculated by an integration process over the various panels and sections. This method relies on the basic hypothesis that total net drag is equal to the sum of the drags of its components and that no interference occurs. There is some evidence that, at low angles of attack, the disturbance of water flow by one piece of netting will affect the drag of those pieces of netting following behind. Another problem is to describe accurately the geometry of a full-scale net in terms of simple shapes. The division of the net into simple shapes is governed by the overall shape of the net. This is dependent not only on the hydrodynamic forces on the netting but also on the associated net rigging—otter boards, bridle system, weights, floats, etc. Separate tests are therefore needed to ascertain the overall shape of the full-scale net.

(ii) The second technique used in model tests is to keep the Froude number constant. This keeps the ratio of dynamic to static forces the same for the model as for the full-size nets (Dickson, 1961; Arlotte and Crewe, 1965). In practice this means ensuring that the flotation to drag ratio remains constant. To do this, whilst reducing all dimensions of length to scale, requires that the towing speed must be reduced by the square of the scale (Dickson, 1961). This applies to towing tank tests with models of actual nets (as opposed to panels of netting).

It is immediately obvious that the requirements of the two techniques are incompatible. To reduce the towing speed by the square of the scale means reducing the Reynolds number.

The basic formula for the drag of a body is

$$\text{Drag} = C_D \tfrac{1}{2}\rho V^2 A,$$

where

$$\rho = \text{fluid density},$$
$$V = \text{fluid velocity},$$
$$A = \text{surface area of the body} \quad \text{and}$$
$$C_D = \text{coefficient of drag}.$$

Now C_D, which is usually determined experimentally, is a function of the Reynolds number. Thus success in the determination of net drag by technique (ii) relies on C_D remaining approximately constant over the relevant region of changing Reynolds numbers. Unfortunately, not enough is known of the behaviour of large areas of netting over a wide range of scale and speeds. If netting is regarded as a combination of cylinders (= lengths of twine) and

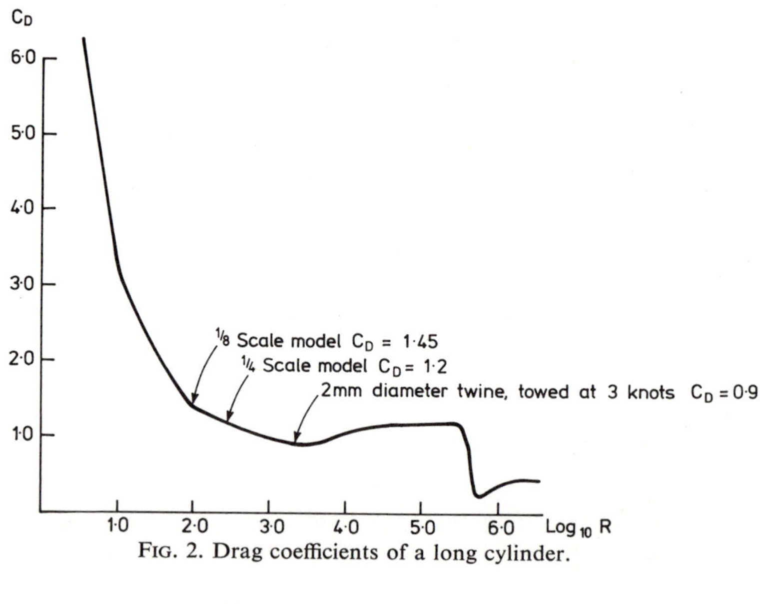

FIG. 2. Drag coefficients of a long cylinder.

spheres (= floats and knots in the twine) then we may look at the appropriate graphs of C_D against Reynolds number for such shapes (Fig. 2) where it is clear that small-scale models of netting are likely to produce misleading results. The corresponding graph for a sphere has a similar shape. However, in the case of floats the Reynolds number value is dangerously close to the critical region where flow becomes turbulent and C_D drops dramatically. Thus experimental modelling of a net complete with floats and weights may again produce misleading results.

Whatever technique of modelling is applied, there are considerable problems and a fair amount of theoretical "hand-waving" is necessary to use experimental results for full-scale predictions. In consequence, the agreement achieved is not very encouraging when comparing these predictions with full-scale tests performed at sea.

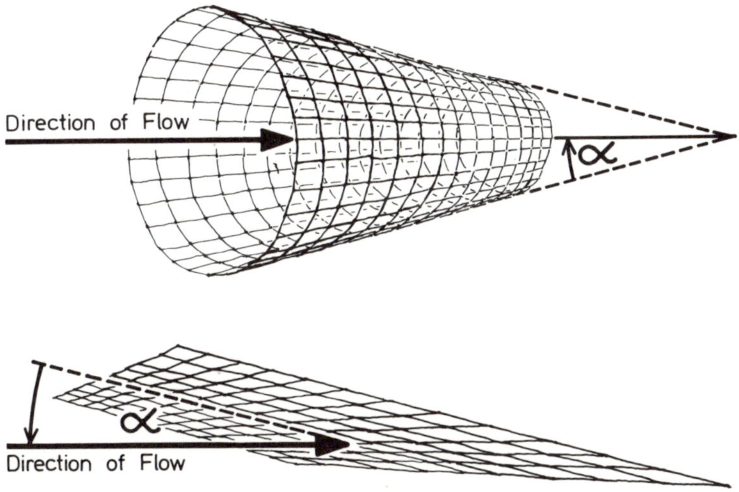

FIG. 3. Different-shaped sections of netting having the same angle of attack to the direction of flow.

A specific example of model tests and theoretical hand-waving (Fridman and Dvernik, 1973) proposes that, to a first approximation, the drag of a piece of netting depends only on the netting area and its orientation to the direction of flow. Thus a flat piece of netting of area A, set at an angle of α° to the direction of flow would produce the same drag as the same area of netting formed into the section of a circular cone with apex angle α (Fig. 3).

The authors propose an elaborate scheme for calculating α_m, the mean angle of attack of a net and then perform experiments on simple models to determine the relationship between C_D, the coefficient of drag, and α_m. Their experimental results show a minimal amount of scatter. To test the applicability of this theory to full-scale nets, three different pelagic nets were selected on which engineering trials have been done. Using the measurements of drag, speed, headline height and wing-end spread it was possible to calculate α_m, using the formula proposed by Fridman and Dvernik. $C_D = \text{drag}/\frac{1}{2}\rho v^2 A$ was also calculated and the resulting values of C_D versus α_m were plotted to compare our practical results with the model tests (Fig. 4). It is immediately obvious that our full-scale nets are not behaving in the same way as those models tested by Fridman and Dvernik.

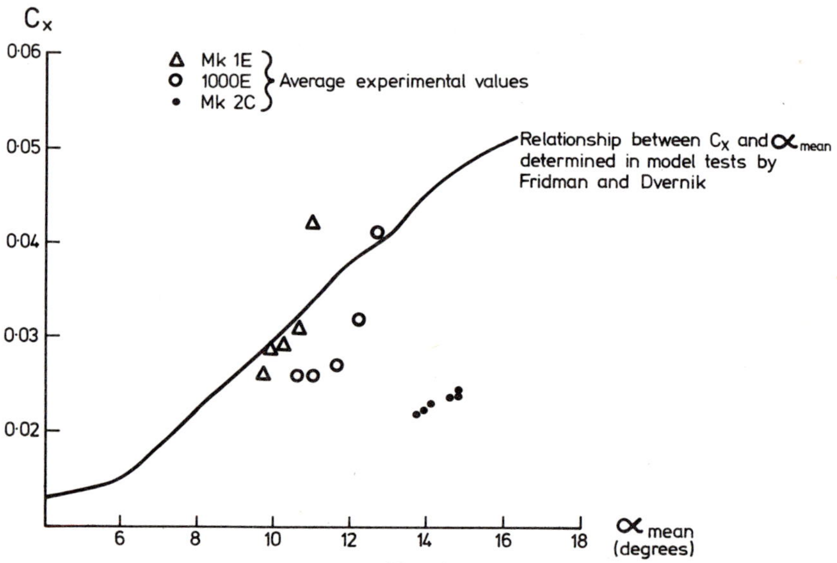

FIG. 4

We have approached the problem of describing net drag in a different way. Rather than fit theory to experimental results from model tests we sought to make a mathematical model that would fit our full-scale trials data. This method consisted of examining whether various formulae would fit our experimental curves, the *format* of these formulae being derived from previous theoretical work. Most successful was a formula

$$\text{Drag} = \frac{K_1}{K_2 V + 1} V^2 A$$

where the term $K_1/(K_2 V+1)$ models the relationship between the coefficient of drag, C_D, and the mean angle of attack of the netting. It was found that the use of the same constants K_1 and K_2 worked well for a large range in size of pelagic nets (Fig. 5). These nets (types 1000E, MK1E, MK2C) have

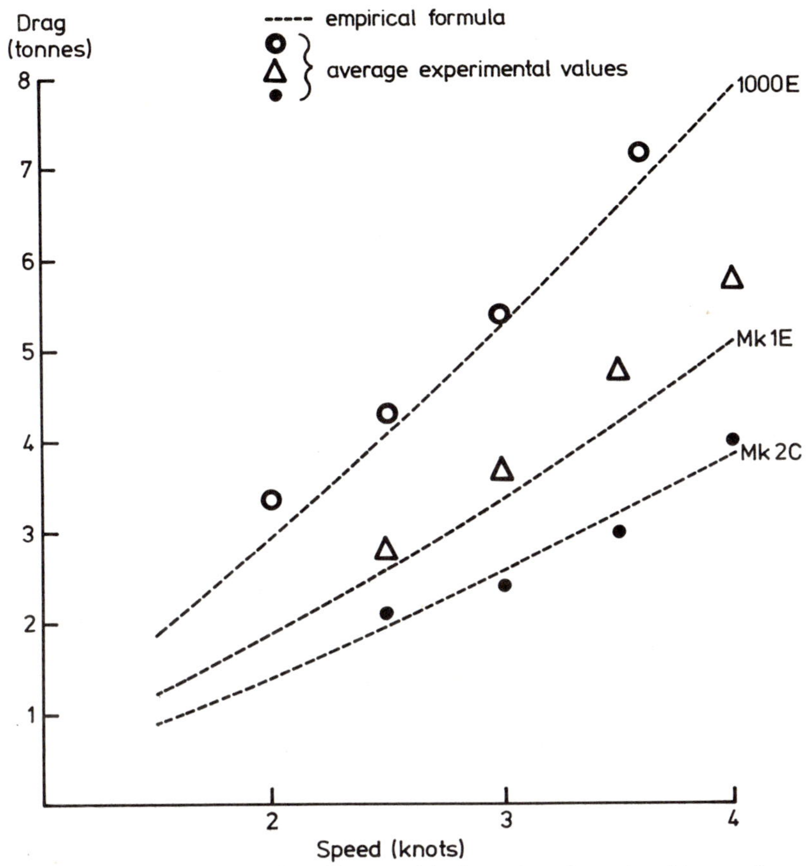

FIG. 5. Comparison of drag predicted by empirical net drag formula with experimental drag values.

similar shape and all are made from the same material, but are of different sizes.

Theoretical modelling to assist in practical fishing gear problems

A typical example of the application of the techniques described concerns a fishery for blue whiting. Two pair trawlers (26 m in length) were chartered to fish down to 500 m (much greater than normal commercial fishing depths). Pair trawling differs from single boat trawling (Fig. 1) in that spread is achieved by the separation of the vessels, avoiding the need for otter boards. Heavy weights are positioned a few metres ahead of the net, and the two legs are usually extended all the way to the ship. It was necessary to predict the warp lengths and additional weights required for the greater depths.

The warp size, vessel characteristics and net specification were available but could not be changed before the start of the exercise. Initial calculations were made for (a) net drag for a range of speeds and (b) achievable towing speed from vessel power. The computation of the wire shapes has been standardized and appropriate computer programs developed.

For this particular problem the tension vector at the warp top end is not known. The available quantities at this stage are net drag and gear weight which provide two components of the tension at the warp lower end. The third quantity is obtained by assuming the horizontal geometry is defined by the practical ship separation distance and net spread. It should be noted that the geometry in the horizontal plane has only a second-order effect on the important vertical geometry (see the following discussion of sensitivity).

The computation is then carried out for a range of towing speeds, warp lengths and gear weights, and suitable values recommended to the vessel operators. In this case, it was to carry 1800 m of warp together with 525 kg weights added to the rigging. This was predicted to allow the vessel to tow at 4 knots and fish the gear down to 500 m.

Sensitivity analysis techniques were applied to the calculations. Each of the main listed parameters were separately increased by 10% producing the following percentage errors in net depth (+ve indicates increase in depth);

Warp circumference	−0·2
Warp weight/unit length	3·7
Hydrodynamic coefficient K_1	negligible
Hydrodynamic coefficient K_2	0·2
Net spread	negligible
Ship separation	−0·3
Warp length	14·1
Net drag	−8·0
Towing speed	−0·6
Gear weight	5·3

Note the minimal dependence of net depth on horizontal geometry net spread and ship separation. Choosing more realistic possible errors in the input parameters, i.e. 10% in net drag value, 5% in gear weight and 3% in warp length, and combining these statistically, gives a range for calculated net depth of ±11%.

Once the vessels were fishing limited measurements were made which were in good agreement with prediction. At 3 knots extremely good correlation was found, but at higher speeds there was a tendency for the gear to reach slightly greater depths than predicted (maximum 5%). For this application high precision over a large speed range was not so important as being able to determine quickly and in advance of the fishing exercise, winch modifications, warp lengths and appropriate weightings.

6. Analysis of Gear Performance

The development of the models discussed in the previous sections has proved to be a long-term project. Until such techniques are available for assessing gear performance a substitute method is required to compare the performance of different trawl gears so that it is possible to judge when a "better" gear has been designed. For this it is necessary to establish relatively simple criteria for such judgements. As far as the fisherman is concerned, it is simply a financial question of maximum return for minimum outlay, but this depends on many variables such as market conditions, labour availability and other factors outside the immediate scope of gear technology and research.

A simpler criterion within the scope of gear research would be the amount of fish caught per unit time. Comparative fishing trials are a recognized method of achieving this end by fishing two gears alternately from the same ship, or from two similar adjacent ships. Statistical analysis of the catch composition of each pair of hauls provides a comparison of the performance of the two nets. Because of the great variability in catches a large number of hauls is needed. Thus comparative fishing is a costly exercise. It does not give an absolute measure of performance, but is only relative to another gear at a given time and place. Lastly it cannot reveal the underlying principles governing the behaviour of the trawl and its ability to catch fish.

The method used at this laboratory for performance assessment is to undertake "engineering" trials. A comprehensive set of measuring instruments is distributed throughout the gear at strategic points to monitor continuously the geometry of, and loading in, a gear or its individual components. These measurements make possible a direct quantitative assessment of the engineering performance of the gear, and, because the environmental conditions defined by water depth, tide and weather conditions are also quantifiable (unlike fish distribution in the case of comparative fishing experiments), then the measure of performance may be said to be absolute rather than relative. Furthermore, such comprehensive data will help to elucidate the principles of the mechanics of trawl gear and will help to supply the empirical constants required for the models described earlier. However, such trials do not determine the catching ability of a net and there is still a place for comparative fishing trials when initial engineering development has been completed.

UNIV. COLL. N.W. BANGOR

Figure 6 shows a typical selection of measurements which is sufficient to define the geometry of the gear and hence to allow resolution of the wire tension measurements in appropriate directions to obtain the drags, spreading forces and lift forces generated by each component of the gear. In some cases measurements of the warp tension at the ship, warp spread at some point above the otter boards and otter board depth may be the only measurements available forward of the boards. Such information is sufficient to determine the complete warp shape in three dimensions assuming only values of skin friction and pressure coefficients of the wires. Step-by-step integration down the warp, together with a simple iterative process similar to that already described, is used in the general case to give the tension

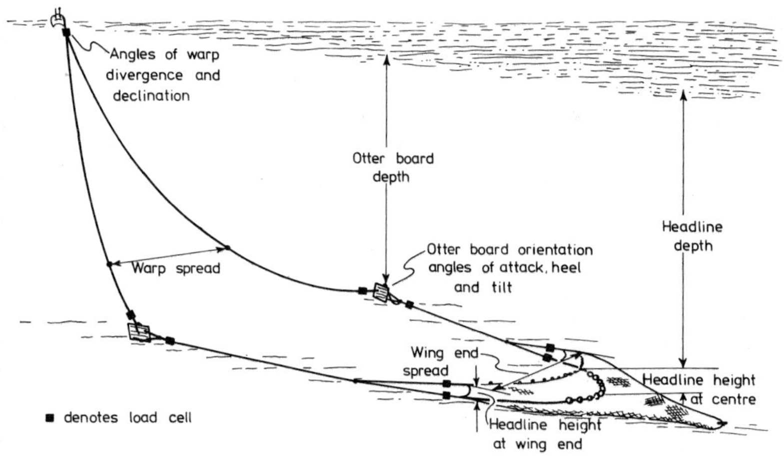

FIG. 6. Some measurements which are made on trawl gear.

components at the lower end of the warp and the spread between the otter boards. Individual computer programs have been written for each combination of measurements so that as much use as possible is made of the data. These programs output the basic quantities used in judging performance.

Three typical examples of such parameters are the following.

(i) *Sea bed area covered per unit towing pull*

Considering the gear as a whole, if the product of effective gear width and towing speed per unit towing pull is increased then an improvement in performance is established. The gear width can be defined as the otter board spread provided that the associated bridle angle of attack does not exceed a specified value. Higher bridle angles will reduce the efficiency of fish shepherding and in the extreme case the net spread will become the effective gear width.

(ii) *Net mouth area per unit net drag*

Considering the net only, if its mouth area per unit net drag is increased then an improvement in performance is established. The mouth area can be defined as the product of headline height at the centre and wing end spread (Fig. 6).

Figure 7 shows a plot of this parameter against towing speed for an experimental net designed at the laboratory, compared with a net used commercially. The improvement in performance is marked although the increase in drag and reduction in mouth area as speed increases clearly shows the penalty of towing a gear too fast. This aspect is important in the problem of matching gear size to available vessel towing power.

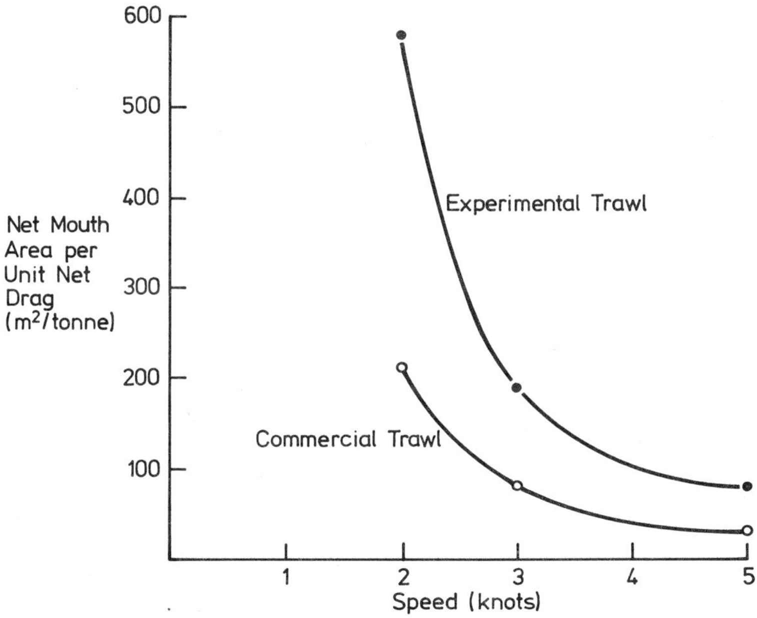

FIG. 7. Performance comparison of two trawl gears.

(iii) *Ratio of force used to spread the net to total board spreading force*

Otter board spreading force is the parameter which determines the otter board and net spreads. However, this spreading force has two components; one is transmitted through the sweeps and is the force used to spread the net, while the other is used to counteract the inpull of the warps. The ratio of the sweep component to the total spreading force generated by the board will vary with warp length and also with size of otter board. Considerable advantage can be gained by using the most efficient warp length for the particular size of otter boards (Fig. 8). If too small a warp length is used then

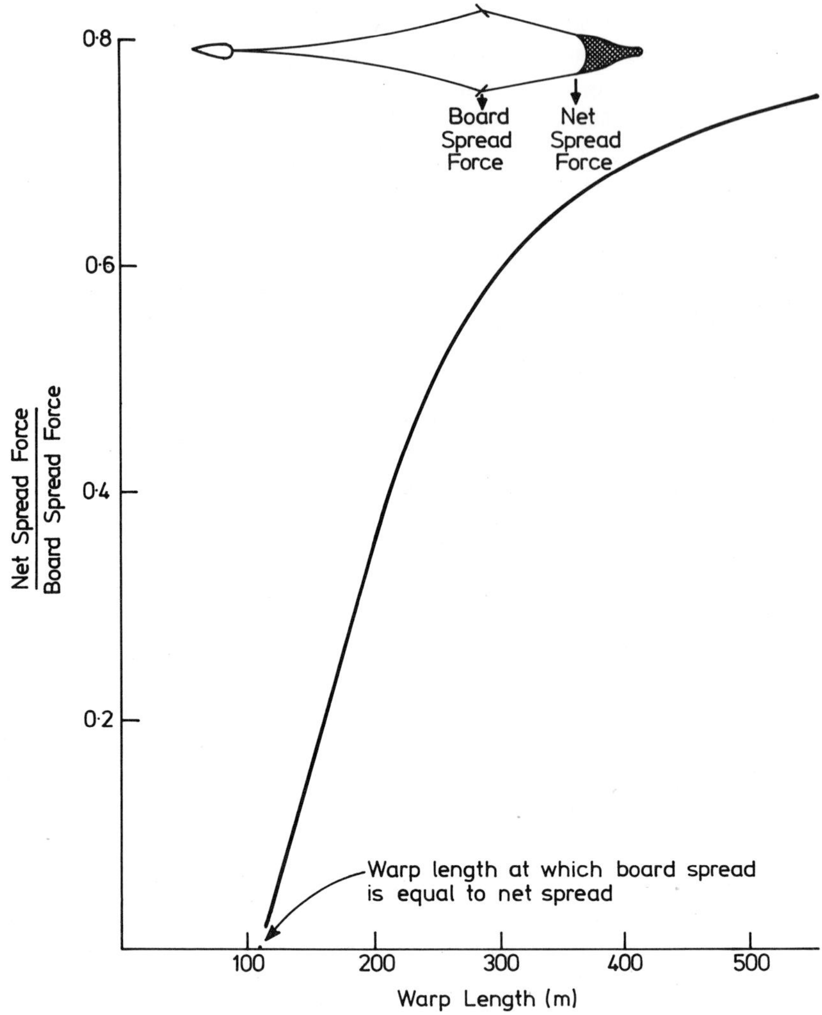

FIG. 8. Optimization of performance of one trawl gear.

only a small fraction of the otter board spreading force will be used to open the net horizontally.

7. Direct Observation and Analysis of Film Records

Another method of monitoring gear performance is by direct observation using divers, television and filming techniques. Direct observation is particularly useful for the determination of net performance in greater detail than is supplied by the engineering trials.

The total drag and the shape of the mouth of the net do not tell enough about the shape and stress distribution in the large area of netting following behind. This part of the gear may have a considerable influence on the overall efficiency, not only because it guides the fish down to the cod-end where they are eventually trapped, but also because high stress areas in the netting will inevitably lead to breakage and the wasting of valuable fishing time in mending.

A team of specially trained divers carries out this observation, working in relatively shallow inshore waters and at towing speeds of up to 3 knots. Each diver records his visual impressions of the net and any measurements of net geometry that he is able to make whilst examining the gear. The divers also use television and cine cameras to record the gear performance on film for subsequent analysis by the gear technologists. Analysis of film and videotape records result, for example, in areas of poorly shaped netting being tailored to minimize stress concentrations and to reduce the possibilities of fish escape. This approach is used in fish behaviour research to study the reaction of fish to the fishing gear.

8. Conclusion

Various analytical techniques in fishing gear research are being developed, each contributing to the determination of the basic principles required to build a model of the fishing gear.

The purely theoretical approach to modelling has been found impractical and has been replaced by a system combining theory with empirical constants and formulae derived from full-scale and tank trials. Consequently the gear model under development may be unsuitable for new gears containing radical innovation without recourse to at least some experimental trials.

There are encouraging signs that a representative model is feasible judging by the success in solving specific prediction problems. In particular, net and otter board models of the simple case where there is no ground contact are satisfactory. It is intended to extend the scope of existing models, particularly in relation to the behavioural interaction between fish and gear components. Programmes involving the tracking of electronically tagged fish in the vicinity of gear, and direct observation of fish reaction to gear, are being carried out.

References

ARLOTTE, T. and CREWE, P. R. (1965). Collected data on model trawl and component tests conducted at Saunders Roe Division. Westland Aircraft Ltd., Report No. H/O/192.

CREWE, P. R. (1975). Trawl performance prediction studies (personal communication).

Dickson, W. (1961). Trawl performance. A study relating models to commercial trawls. *DAFS Marine Research* **1**.

Engvall, L. O. and Engstrom, J. (1969). A method for selection of an optimum fishing vessel for investment purposes. *F.A.O. Fish. Rep.* **83**.

F.A.O. (1975). Expert consultation on quantitative analysis in fishery industries development. *F.A.O. Fish. Rep.* **167**, Vol. 1.

Foster, J. J. (1969). The influence of fish behaviour on trawl design with reference to mathematical interpretations of observations on the swimming speeds of fish. *F.A.O. Fish. Rep.* **62**, Vol. 3.

Fridman, A. L. (1973). Theory and design of commercial fishing gear. Translated from Russian (PPB), U.S. Dept. of Commerce, National Technical Information Service.

Fridman, A. L. and Dvernik, A. V. (1973). Development of a method for the calculation of the resistance of a trawl net. *Fischerei Forschung*, Vol. II, part 2, pp. 7–13.

Hamilton, G. F. (1970). Flow into wedges and cones with porous walls. PhD Thesis, University of Toronto.

Kowalski, T. and Giannotti, J. (1974). Calculations of fishing net drag. University of Rhode Island Marine Technical Reports Nos. 15, 16.

8. The Development and Application of Computer Aided Cartography in the Fishing Industry

J. B. RICHARDSON

White Fish Authority, Industrial Development Unit, Hull

1. Introduction

Until the introduction of electronic aids in this century, navigation on the high seas was a distinctly hazardous task. With the possible exception of coastal waters in sight of land, the only predictable objects for position fixing were in the heavens—the sun, moon and stars. Sea navigation therefore relied to a small extent on technical expertise and to a large extent on pure luck and experience. Obviously in order to reduce the element of chance various charts were constructed over the years but on a rather subjective basis, until the naval hydrographic authorities began surveying areas specifically for the production of navigation charts.

The main interest of these authorities, however, was to chart coastal waters, estuaries, etc., where the avoidance of land was the major problem of the seaman. In deeper water only large wrecks, sandbanks and rocks, etc. were shown on the charts as being particular hazards to surface navigation. Since the second world war, electronic aids to navigation, such as Decca and Loran, have reduced the luck in navigation in most parts of the world to a more acceptable level, but the fact still remains that most charts are constructed to assist in navigation and avoiding hazards on the surface of the sea.

Whilst this is of importance to the fisherman as much as any sea-goer, his problem is much more complex, since his major requirements, and livelihood, lie beneath the surface of the sea. The majority of fishing by the U.K. fleet takes place with the trawling gear on or very near the sea bed, and so the fisherman needs to know more detail of depth, shape, nature of the bottom, etc., as well as the position on the sea bed of hazards to his trawl. For example, a large boulder would cause no hazard to navigation, in a suitable depth of water, but, if caught by a fishing trawl, would not only cause damage or loss to the trawl but probably loss of the catch and hence income to the trawler skipper.

A major disadvantage of the charts produced by the hydrographic authorities is therefore the lack of detailed depth and sea bed information in the areas most widely used by the trawler fleet. A second disadvantage is that of scale. Again, most charts are compiled for navigation over large distances, and therefore produced at small scales such as 1 : 1 000 000 or smaller. Since a typical time for towing a trawl might be 3 hours at 5 knots, covering a distance of 15 nautical miles, it would be very difficult to plot the course in detail on such a chart, requiring a line of approximately 1 inch at a scale of 1 : 1 000 000. In particular if the skipper wished to repeat the tow, the chart would be of little use in assisting him.

It is true that Admiralty charts exist in coastal areas at scales of 1 : 100 000 and better, but these are only of major use to the inshore fisherman, and even then the depth and sea bed data are not always sufficient for his requirements.

2. The Development of Specialized Charts

Prior to the early twentieth century little was achieved in charting to assist the fisherman. The usual method employed was for the fishing skipper to construct his own charts of particular areas, mostly from data he had collected through years of fishing in those areas.

The first major production of specialized charts for fishing was by Messrs. Imray, Laurie, Norie and Wilson, with a series of charts for the British Isles, North Sea, Iceland and the Norway coast. These charts gave much more detailed depth and bottom data than any other type of chart and have been in wide use by the fishing fleet since their introduction. Figure 1 shows a section of one such chart for the North Sea. The scales ranged from 1 : 500 000 down to 1 : 1 000 000 which were still too small to allow the pinpointing of fishing obstructions and to show bottom contours in sufficient detail.

Skipper Tess Johnson, a leading trawler skipper with the Ross trawler fleet, made his name fishing the plaice grounds between the North Cape of Norway and the Russian Murmansk coast. He appreciated the importance of charting his fishing grounds accurately and for 20 years kept detailed records of his best tows and obstructions encountered on the sea bed. On leaving the sea in 1963 he conceived the idea for a series of charts covering areas of particular interest to the fisherman and at sufficiently large scale for trawling. A small company was formed in Grimsby, with financial support from the White Fish Authority, to produce the now famous Kingfisher Charts, the first chart made specifically to suit the trawler skippers'

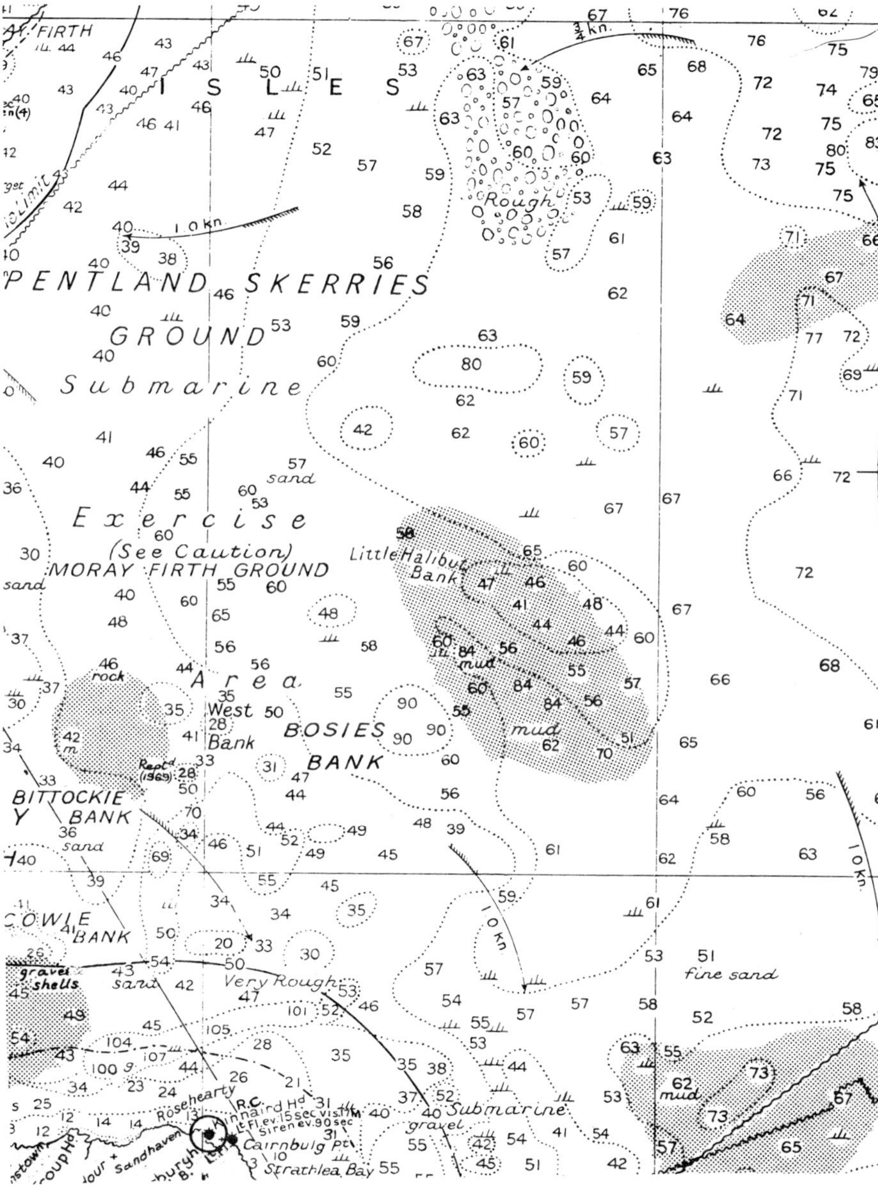
FIRTH
I S L E S
Rough
PENTLAND SKERRIES
GROUND
Submarine
Exercise
(See Caution)
MORAY FIRTH GROUND
Area
sand
Little Halibut Bank
mud
rock
West Bank
BOSIES BANK
BITTOCKIE BANK
COWIE BANK
gravel shells
Very Rough
fine sand
Rosehearty
Kinnaird Hd
Cairnbulg Pt
Strathlea Bay
Submarine gravel
1 0 kn
3/4 kn

FIG. 1

requirements. The first chart contained obstruction and sea bed contours taken solely from Skipper Johnson's own records, but this soon stimulated other trawler skippers to provide similar data of their own.

Over the next 10 years the Company produced about 60 charts of major deep-sea fishing areas, at scales of between 1 : 100 000 and 1 : 200 000 together with nearly 20 inshore charts at scales as large as 1 : 50 000. A section of a typical chart is shown as Fig. 2, on which can be seen fishing obstructions represented by small circles containing a number. The number refers to a table printed down the side of the chart, containing the position as reported, for example, as a two-lane Decca reference. This system not only allows the skipper to plan a tow on the chart, avoiding any obstructions, but also to make direct comparison between his Decca receiver readings and the Decca position shown in the table, whilst actually executing the tow.

3. The Advent of Automation

The changing demands of the fishing industry, in terms of alternative fishing areas, exploitation of more distant grounds, etc. eventually required the increase of new chart production beyond the means of the small Company. In addition, the constant necessity of updating existing charts with new obstruction information, changed fishing limits, etc. became an overwhelming burden for the already overworked staff. Consequently in 1971 the production of Kingfisher Charts was taken over by the Authority in Hull, but still controlled by Skipper Johnson. As the Authority's Computer Section is housed in Hull, this provided an ideal opportunity for investigating areas in which the computer could be applied to increase efficiency and accuracy, and reduce the manual effort required in producing the charts.

The single most laborious job manually was in the addition of a Decca navigation lattice. This involved plotting lane points on the chart using tables of Decca and latitude/longitude positions, and then using splines, etc. to produce a smooth lane. Typically a Decca two-lane pattern might require 150 lanes plotting on a chart: an awesome job for the draughtsman. The first step in computerization was therefore to design a method of computing the positions of these lane points, and constructing a smooth curve through them. With a Ferranti master plotter it was then possible to "draw" the lanes on a photographic film.

In order to apply this system in producing a chart, it was necessary to obtain an existing chart containing the area to be latticed. The appropriate section of this chart was then photographed and enlarged to the required scale. This process, however, introduced distortion, which was evaluated by accurately measuring the side-lengths of the photograph and converting

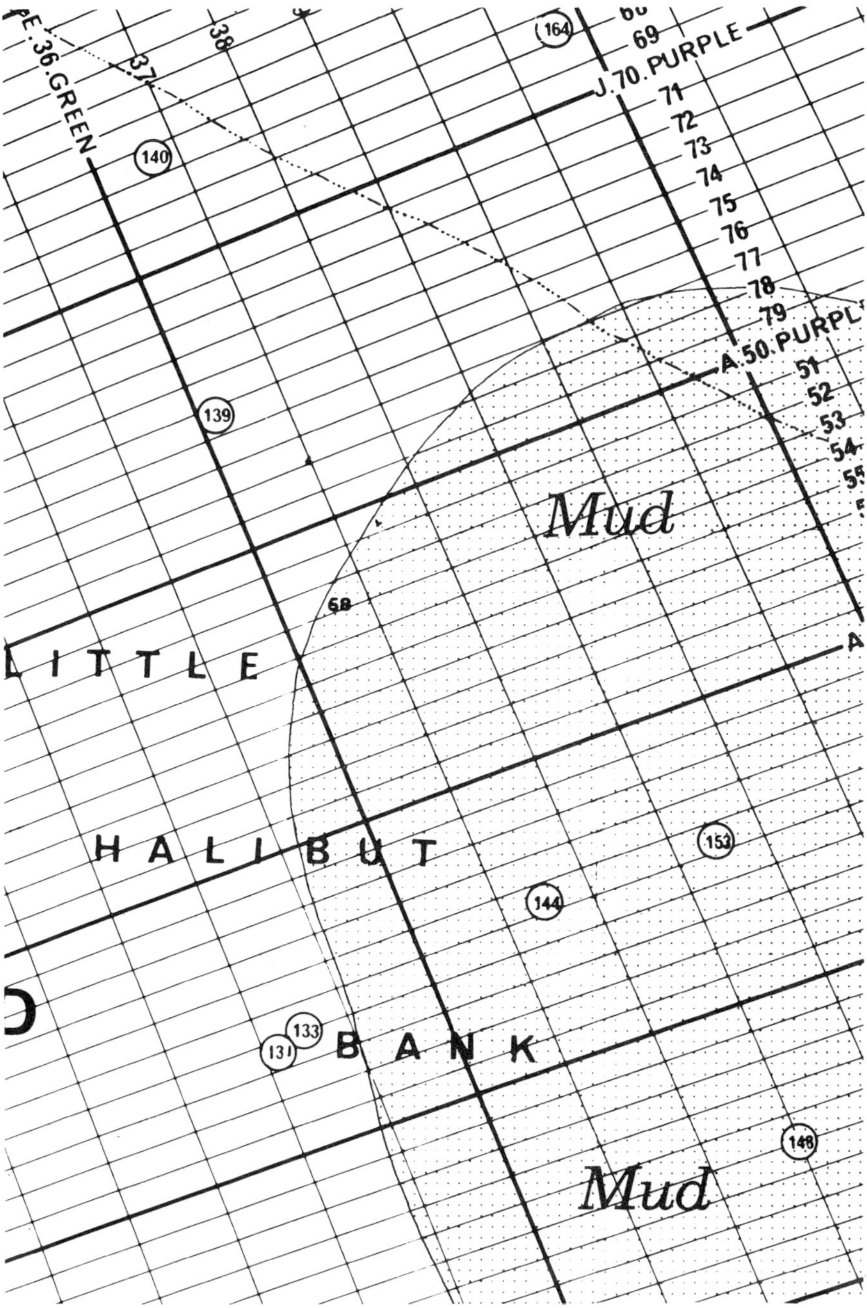
E.36.GREEN
37
38
164
69
J.70.PURPLE
71
72
73
74
75
76
77
78
79
A.50.PURPL
51
52
53
54
140
139
Mud
LITTLE
HALIBUT
BANK
153
144
133
131
148
Mud

FIG. 2

these, through the standard formulae for Mercator's projection, into X and Y scales for the chart. By using independent scales in the two directions it was then possible to inject a similar amount of distortion into the computer-produced lattice and so obtain a lattice overlap of exactly the same dimensions as the original photograph. A similar process was used to produce a standard Mercator border around the lattice, so giving absolute geographic reference to the projection. The computer-produced film was then simply placed over the enlargement and all the chart data traced by hand on to it.

The process obviously does not produce a true orthomorphic Mercator projection, but, since there is equal distortion in the two components of the new chart, the computer-produced film and the enlargement, the result is still accurate; certainly as accurate as any hydrographic chart printed on paper. This system has been successfully reducing the manual effort required in chart production for about 3 years, but it is now being replaced in most areas by the fully automated method, next described.

4. Full Automation

After the success of the latticing system the natural progression was into the full field of computer-aided cartography. This, however, required the purchase and hire of specialized equipment. The major problem in using a computer is that both the raw data and end-product are in the form of charts. There has therefore to be some means of transforming existing chart data, including land and depth contours, into a numeric form for computer processing, and then transferring the processed numbers back into the different contour lines at some future date. The equipment required to do this was supplied by the Information Equipment Group of Ferranti Ltd.

First, the Authority purchased a Freescan Digitizer for use in Hull. This is a draughting table that can accurately sense the position of a cursor moved in contact with its surface, and convert this position into an X–Y coordinate punched on to paper tape. In practice the cursor is moved slowly along the chart contours and the position recorded automatically at a preset rate, the process being termed "digitizing". Land and depth contours, etc. are thus translated from lines on a chart into a reel of punched paper tape numbers, which can then be processed on the computer.

To reconstitute these point data into a chart a system of computer programs is required that produces a set of plotting commands, stored on magnetic tape. These can be used to drive plotting machinery, available on a bureau basis at the Ferranti offices in Edinburgh. Two types of plotter are available. One is a very fast Microplotter, producing a 35 mm type aperture card.

This is used in the initial stages of checking the digitizing, where speed and cheapness of service are required. However, the checking of the finer details of the chart is not possible at this size so a full-sized plot must be produced on a Master Plotter. This has two modes of operation, either with a ball-point pen on paper, or photographically on film. As the latter process is expensive, it is only employed to produce the finished "master" chart, which can then be used for printing copies of the Kingfisher Chart for sale.

All computer processing is done in Hull on the Authority's ICL 1901A computer and the programs to operate the system were designed and written by Authority personnel within the computer section. The full system has been in use for most of this year and used extensively in the recent series of charts covering the North Sea.

5. General Principles

The main aim of the automated system is to produce Kingfisher Charts, with as little manual intervention as possible, from data digitized from other charts of differing shapes, sizes and scales. The process falls naturally into the following stages:

- Stage 1: the chart is digitized and the data edited and plotted to produce a corrected version on magnetic tape.
- Stage 2: the magnetic tape is further edited and then added to a "chart databank". In addition, fishing obstruction information is collected, checked, converted to the appropriate format and added to an "obstruction databank".
- Stage 3: new charts or "windows" are produced for any area covered by the databanks, to any required scale, and with a navigation lattice superimposed.
- Stage 4: the new chart master is completed manually by the addition of lattice numbering, topographical shading, etc., and copies produced for sale to the fishing fleet.

6. Digitizing Process

The initial phase, depicted in Fig. 3, is concerned primarily with data capture, i.e. the translation of existing charts into a corrected computer-processable form on magnetic tape. A chart is placed on the draughting table and aligned with the table axes to within a tolerance of about 0·010 inch. In all cases the datum is set to the bottom left-hand corner of the chart, so that all data are recorded as a series of X–Y measurements relative to a fixed latitude and longitude position.

The information contained on the chart is divided into five distinct types and each is digitized separately in the following order:

land contours,
fishing limits,
depth contours,
miscellaneous contours, bad ground, pipelines, etc.
discrete data including depth soundings, wrecks, etc.

Within a type, the data can be subdivided into any number of sets, each containing a preplanned section of the digitizing. Throughout the computer processing all contour data within a set are considered to be continuous, and when plotted are reconstructed by joining each point to the next by a straight line. Data in different sets are considered to be independent and so contours within different sets are not normally joined together although

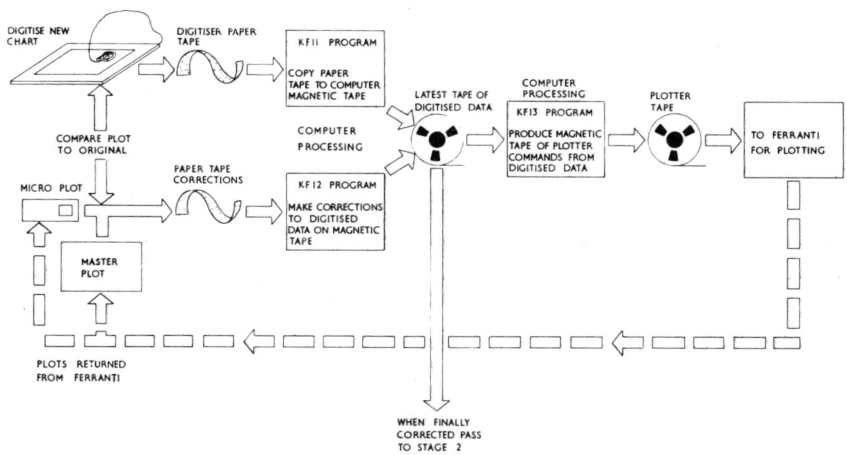

FIG. 3. Stage 1 flow diagram—digitizing a chart.

circumstances will be described later where this could happen. Using this simple technique the operator can take account not only of islands and closed contours, but also plan the work on lengthy contours into short runs of more manageable size. A keyboard, attached to the digitizer, allows start and end codes to be entered into the digitizing stream to show the limits of each type or set.

The digitizing process itself is simply a matter of using the cursor to trace contours on the chart. By setting an appropriate time interval on the control unit, readings are taken automatically as the cursor is moved. In this way, the pictorial chart information is translated into punched paper tape co-ordinate data for computer processing.

The first computer process is in fact to copy the paper tape on to magnetic tape, using program KF11, mainly for ease and speed of processing. At the same time a printed version of the digitized data is produced to allow the operator to check and edit the data. At this stage, however, only the codes can be checked accurately. To check the detailed digitizing requires plotting of the data via program KF13, which will convert any digitizer tape into a tape of plotter commands for processing at Ferranti.

As stated previously, plots are produced on the Microplotter in the initial stages of checking, providing a fast, cheap means of detecting major errors in the work. Final detailed checking requires a full-sized master plot, produced on translucent paper, which can then be superimposed on the original chart. Corrections are applied to the digitizer tape using program KF12, which allows the addition or deletion of any group of points or codes on the tape. After completing the correction/plotting cycle as many times as required, the end-product is a corrected digitizer tape available for further processing in the next stage.

7. Formation of the two Databanks

Long-term storage of the data for new chart production is on two databanks, stored in magnetic tape. The first of these, the "chart databank", holds all data from the digitizer tapes produced in Stage 1. However, before these data can be added, certain inaccuracies must be checked. First, the data required for the databank must be unique, and so any sections overlapping with other charts, no matter how small, must be removed. Secondly, in order to ensure smooth continuation of contours, between adjoining chart areas, it is necessary to obtain the boundary intersection point for each contour digitized. Finally, any distortion of the original chart must be removed to reinstate an orthomorphic Mercator projection. Program KF21, as shown in Fig. 4, in addition to other code and format-checking facilities, provides these

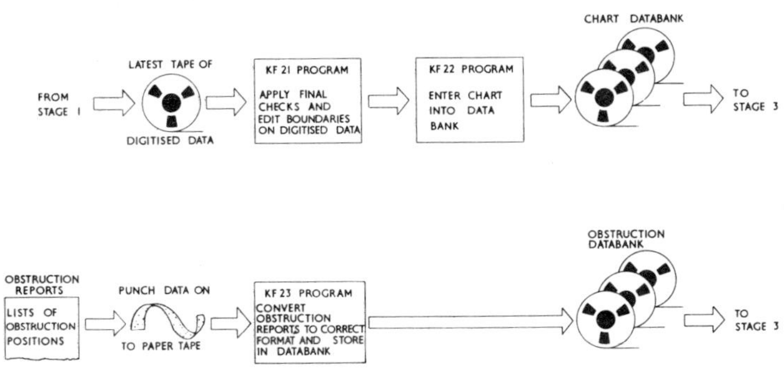

FIG. 4. Stage 2 flow diagram—formation of databanks.

functions and produces a unique set of chart data for storage in the chart databank.

In theory one large databank could be used for holding charts of all areas, but in practice several are used, not only to reduce the problem to a more manageable size, but also to remove some of the problems of differing chart data. A standard earth datum is chosen for each area and any chart not based on this is converted during the running of the program. Program KF22 completes the stage for digitized data by inserting the chart in the appropriate position in the databank. The logic is simply to place the chart behind its nearest neighbour on a magnetic tape provided this is within a given distance, otherwise a new tape is created. This allows the databank to be spread over several magnetic tapes and allows for the minimum number of tapes to be scanned when plotting a new chart.

The second databank created in this stage is independent of the digitized chart data, and concerned mainly with handling obstruction data. Obstruction positions are reported by the fishing fleet in a variety of ways. The majority are given as Decca or Loran coordinates but alternatives are radar range and bearing or even latitude and longitude. Program KF23 provides two functions, by converting the reported positions into latitude and longitude, and then storing the report in the "obstructions databank". In addition, other discrete data such as additional depth soundings, names, can be stored on this databank. This allows charts to be updated by first-hand knowledge from the fishing skippers, without having to digitize small amounts or change the otherwise "permanent" data held on the chart databanks.

8. Production of a New Chart

The system so far has been concerned with the capture of data in a form suitable for the production of new charts. Stage 3, as depicted in Fig. 5, allows the plotting of a Kingfisher Chart from the two databanks described, as a rectangular area to any scale within the physical limits of the Ferranti master plotter. In practice, the large area covered by a databank is subdivided for chart production into a series of rectangular "windows" at scales of between 1 : 100 000 and 1 : 200 000 and measuring approximately 40×30 inches. The process normally requires four computer programs to be run, each forming at least one "plot" on magnetic tape. All plots are then superimposed on to one film on the master plotter, and a master transparency of the chart produced.

The first program, KF31, is supplied simply with the latitude and longitude of the bottom left- and top right-hand corners of the chart area, together with the scale. The program then identifies all charts in the databank that contain digitized data within the bounds of the window, and, for each set of each type of data, produces the appropriate plotter movement commands to reconstruct the contour, etc., at the required scale. Each data type has a particular line thickness and type associated with it, giving for example a 0·025 inch continuous line for land, and a 0·010 inch dashed line for fishing bank boundaries. Where a contour crosses a boundary, the point of intersection is interpolated from the digitized data, so that, where required,

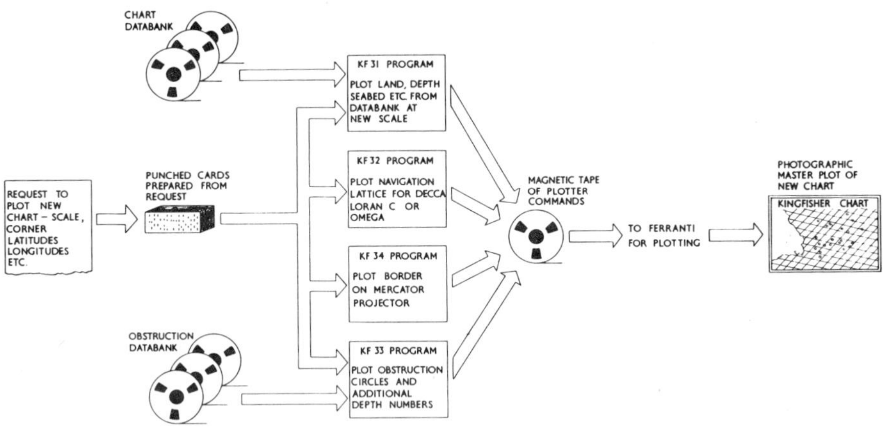

FIG. 5. Stage 3 flow diagram—production of a new chart.

contours stop exactly on the border. If a contour ends inside the window, the end-point is stored and, whenever a contour of the same type is plotted from another databank chart, an attempt is made to join the two, provided the end-points lie within a given distance. In this way small gaps in the digitizing, at existing chart boundaries, are corrected and the result made to appear continuous. At present all contour plotting is done by joining digitized points by straight lines. Provided the digitizing has produced a sufficient number of points, and provided the scale change required is not too large the technique is quite satisfactory. However, there may be a requirement in the future to introduce some form of interpolation and smoothing into the plotting procedure.

All new Kingfisher Charts now carry a navigation lattice, produced by program KF32, which can currently handle the Decca, Omega and Loran C systems. The program requires no digitized data but computes lane points from the geometry of the earth and the navigation system parameters. Each lane, although curved, is again plotted as a series of straight-line segments. The points are selected by an iterative process, such that, by joining any two consecutive points by a straight line, the deviation from the curved lane lies within a given tolerance. The program automatically tries to increase the length of the straight line at each point until the tolerance is exceeded, when the last successful point is used for plotting purposes.

The addition of obstruction data is via program KF33, which selects obstruction reports from the appropriate databank area and plots these as small circles around the position of the obstruction. A number plotted inside the circle gives a unique reference for this chart of all obstructions plotted. A table of reference numbers against obstruction positions, usually relative to a navigation system, is automatically produced. This is later printed down the side of the chart to give a direct position reading for the fishing skipper.

The final process in plotting the chart master is the construction of a Mercator border to surround the windows, using program KF34. This produces a standard chart border, complete with grid ticks, corner markings and latitude and longitude references.

9. The Final Process

The result of Stage 3 is a magnetic tape of plotter commands which is then sent to Ferranti, where a photographic master plot of the chart is produced. This, however, still requires certain additions which must be done by hand, because they are either not economical or not possible to produce automatically.

Since the navigation lattice is produced independently, at the time of the computer run it is not possible to predict the position of other chart data. Therefore not only does the lattice cover any land on the chart, but the lane numbers cannot be plotted since this could well obliterate more important information. To improve the appearance of the chart any lattice lines are removed from land areas and the lane numbers added in the appropriate positions. Similarly, any shading required to show either topographical features or indicate bad fishing ground etc., must be added manually using "Letratone" or similar.

Finally, the system does not cater for varying name size so that it is not possible to plot large-size bank or island names and small-size lighthouse references, etc. All this detail is, however, digitized, but plotted at the smallest size ever required. Whenever an alternative is needed, this must be added in "Letraset" using the plotted name as a position reminder only. In this way the greatest flexibility is maintained without having to rely on the draughtsman to remember the position of the data. A day's work on the master chart, after plotting is complete, can enhance the quality of the product to maintain the high standard required of a Kingfisher Chart. The final production of chart copies is by a standard dyeline printer. The main advantages of this are the cheapness and the ability to produce copies on demand. This allows the charting section to maintain a library of masters, which can easily be kept updated, without the problems of large volume storage required if many copies are produced. The main disadvantage is that the process results in a black-and-white copy which removes the benefits of contrasting colours. However, since colour was originally used mainly in the navigation lattice, using suitable line thicknesses in plotting and labelling lanes more frequently keeps the disadvantage to a minimum.

10. Conclusions and Future Enhancements

The fully automated system of chart production is now complete and in daily use, although by no means to the full extent of its capabilities. The major problem at present is in the completion of Stages 1 and 2 for sufficient charts to produce a reasonable-sized databank. Digitizing and the completion of Stage 1 for an average chart can take up to 10 weeks. It is therefore necessary to plan areas to be digitized well in advance of production requirements for new charts. Once, however, a databank of sufficient size has been created, the average time to produce a new master chart is approximately 1 week and the manual effort required reduced by about 90%.

The over-riding justification for using the computer has been the necessity of keeping pace with the changing face of the fishing industry and the constant demands for new charts and wider coverage. Within the past 5 years it became obvious that manual production, without a large increase in labour force, would not be able to satisfy these demands and so automated cartography was the only really effective solution. Economically, if all attributable costs and overheads were accrued directly to chart production, the decision may not have seemed immediately justifiable. However, a major part of these costs would be computer time, and since the computer has spare capacity over the work for which it was purchased, the real overhead of automation is the purchase of a digitizer and the cost of bureau plotting at Ferranti. Compared with the cost of additional staff required to produce the same volume of charts this becomes almost insignificant.

Future work on the system lies in two main areas. Firstly, there is a constant need to increase the efficiency of computer and plotter usage, and hence cut the costs, as well as enhance the quality and detail of the end-product in terms of additional line types, conventions, metrication, etc. Secondly, investigations are in progress to record depth and position automatically whilst fishing, using simple data-logging equipment to record from standard echo sounders and navigation equipment on board most fishing vessels. The data would then be processed in Hull and used for automatic depth contouring to construct more detailed contours for the charts. The progress, however, is relatively slow, since commercial fishing vessels do not make ideal survey vessels when catching fish, and so data from many trips over the same area must be analysed before even simple contours can be constructed. Ultimately, however, more contours will be made available for addition to the databanks, and thence to the next generation of Kingfisher Charts.

9. Mathematical Theories of Fish Swimming

SIR JAMES LIGHTHILL

University of Cambridge

1. Introductory survey

The hydrodynamic interaction between a swimming fish and the water is a good example of a problem where, although a correct mathematical expression of the laws of motion has been known for over a century, any direct numerical calculation of the motion from those laws is far beyond the capacity of any existing computer. The reason lies in the prohibitive storage requirements of a problem where the hydrodynamic equations have four independent variables (three of space and one of time) and there is an enormous ratio between the fish's linear dimensions and the shortest distance over which significant variations of flow occur; namely, the thickness of the boundary layer of water dragged along close to the fish's surface by its frictional action. This, then, is one of the innumerable scientific fields where the role of mathematical theories is to bridge the gulf between a natural phenomenon's natural complexity and the strictly limited capacity of electronic computers.

Benefits to zoology from mathematical theories of fish swimming may include improved understanding of how differences in fish morphology are related to differences in swimming capabilities required in different modes of life. Benefits to fisheries may include improved understanding of the limits to fishes' capabilities for action to avoid capture.

Space permits discussion only of carangiform swimmers (including a majority of commercially interesting fishes but excluding, e.g. eels, dogfishes and flat fishes). They propel themselves primarily by active tail oscillations to which, in large part, the remainder of the body makes passively responding motions.

Perhaps the best kind of mathematical theory of carangiform swimming which can be offered at present is one which treats the hydrodynamics of discrete fins (especially, the caudal fin) along lines pioneered in aeronautics for treating the unsteady aerodynamics of oscillating wings. As regards the elongated body of the fish, two separate hydrodynamic features must be taken into account:

(1) the frictional resistance to its motion, to be estimated through data and theories regarding the boundary layer already mentioned;

(2) the forces of interaction with the water associated with its lateral motions, to be estimated by "elongated-body theory" with its emphasis on the notion of added mass.

These methods are applicable both in steady swimming, where propulsive forces from the tail region must balance body resistance while side-slip and yaw in the anterior part of the body need to be minimized, and in manœuvres where the fish seeks to make a sudden acceleration or change in direction.

For example, the starting and turning of goldfish and rudd were treated in this way by my colleague, Dr. D. Weihs (1972). For their caudal-fin hydrodynamics he used the unsteady aerodynamic theory of "delta-wings" (Lomax, 1962). For the forces associated with motions of the anterior part of the body he used elongated-body theory, together with estimates of frictional resistance.

Some other problems of great interest are posed by the high-aspect-ratio caudal fins of crescent-moon shape ("lunate tails"), which are characteristic of the fast-swimming scombroid fishes (including mackerel, tuna and swordfishes). I had given (Lighthill, 1970) a simple theory of lunate-tail hydrodynamics. This was based on unsteady aerodynamic theory for wings of infinite aspect-ratio making heaving and pitching oscillations of small amplitude. (For fishes, of course, these up-and-down aerofoil motions are turned through 90° to become the lateral oscillations of a caudal fin.)

More recently, my colleagues, Dr. M. G. Chopra and Dr. T. Kambe, used various sophisticated computer programs derived from aeronautical practice to calculate more accurately the hydrodynamics of lunate tails. Chopra (1974) extended my simple theory to wings of finite aspect-ratio, confining himself however to rectangular planforms. Chopra (1976) gave a separate extension to oscillations of large amplitude (although only for wings of infinite aspect-ratio). Finally, Kambe and Chopra (1976) adapted an excellent Royal Aircraft Establishment program due to Davies (1963) to obtain predictions of the hydrodynamical behaviour of a wide variety of different shapes of planform of finite aspect-ratio.

Drs. Chopra and Kambe have used this work to throw light on the optimization processes that can contribute to reducing the rate of working on the lunate tail required to generate a given thrust. They have discussed also in some detail how that thrust is balanced by the hydrodynamic resistance due to body friction and, also, to the respiratory flow through the gills.

As a piece of work complementary to those advanced treatments of caudal-fin hydrodynamics, I want in this paper to discuss what is from some points of view the central problem of carangiform swimming. The caudal fin generates thrust through its lateral oscillations. Those in turn are produced by oscillatory *side-forces* with which the body musculature acts on the caudal fin. To every action, however, there is an equal and opposite reaction. Thus, the anterior part of the body is also subjected to oscillatory side-forces, which are equal and opposite to those with which it acts on the tail. Any large yawing motions, with which the anterior part of the body may respond to such oscillatory side-forces, could greatly increase the energy dissipation involved in swimming.

Lighthill (1970) argued that two characteristic features of the morphology of carangiform swimmers have the function of reducing these "recoil" motions (yawing of the anterior part of the fish). The deep, laterally compressed fish body is important because it responds to side-forces *as if* its inertia were greatly augmented by the "added mass" (also called virtual mass) of water associated with its lateral motions. Yawing response, indeed, is diminished in the presence of a large *moment* of this inertia (including that of the fish mass and that of the substantially greater added mass of water).

Equally important is the presence of a region of greatly *reduced* depth of fish body lying *between* the actively thrusting caudal fin and the passively responding anterior portion. Lighthill (1970) showed that a big increase in the oscillating side-force must result from rapid changes of momentum of the water in this region unless the local depth (and therefore also the local added mass) is greatly reduced. Such a region of greatly reduced body depth immediately proximal to the caudal fin is, in fact, consistently found among carangiform swimmers.

Against this background, I want to describe in Section 2 a new theory of recoil; that is, of those passive yawing motions with which the anterior part of the body responds to the force acting between it and the tail. The theory aims to estimate *both* the amplitude of those yawing motions *and* any associated energy dissipation.

I discuss the common case when the anterior part of the body is relatively inflexible. It acts on the caudal fin with an oscillating side-force, producing in the fin a large amplitude of oscillation. The equal and opposite reaction on the inflexible anterior part of the body makes it oscillate with smaller amplitude (and, actually, different phase) because of its high effective inertia. A flexible link reconciles the two different motions; it also transmits the force between them with little modification, because (as a result of its greatly reduced depth) its own inertia is small and so is the hydrodynamic resistance to its movements. In fact, only the anterior portion, with its large depth, possesses the "purchase" necessary to act on the tail with the lateral forces required while itself "giving" (that is, recoiling) to only a small extent.

I believe that the theory of Section 2 may be of interest because it uses quite simple considerations of inertia and resistance to indicate what determines the amplitude of recoil motions. In addition, it gives quantitative estimates for how great the body depth must be to reduce those recoil motions so that they do not add greatly to the rate of energy dissipation.

In Section 2, then, I describe a theory of the recoil of a relatively inflexible anterior part of the fish body which is flexibly attached to a caudal fin performing carangiform motions. Before doing so, I should go into a little more detail regarding those fin motions.

As Fig. 1 indicates diagrammatically, a caudal fin's carangiform motion involves twists at each extreme of the oscillation so that the lateral movement always includes a backward-facing component. The function of those twists can best be understood if we recognize, first, that *large* lateral velocities for

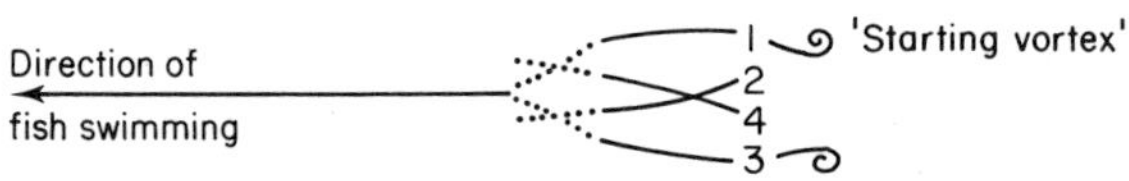

FIG. 1. Illustrating the discussion of the phases of side-forces.

the fin are desirable: they generate a large angle between the direction of swimming and the fin's path through the water, so that hydrodynamic forces at right angles to that path can have a substantial thrust component. Then, however, twist becomes necessary to reduce the fin's angle of attack to that path, so avoiding both flow separation at its leading edge *and* excessive energy wastage in setting up a vortex wake.

There is considerable evidence that the lateral force with which the fish body has to act on a caudal fin in carangiform motion remains nearly in phase with its lateral velocity. This is important because it implies that the fish musculature is not significantly involved in periods of *negative working* (which, as is well known, exact from muscle a substantial positive energy cost). Admittedly, for caudal fins of high aspect-ratio, you might expect lateral force to *lag* in phase behind lateral velocity, through the "Wagner effect"; as the fin's angle of attack to its path through the water builds up, there is a lag (while the "starting vortex" moves away) in the build-up of the force-generating "circulation" towards its full steady-state value. However, another component of the side-force on the caudal fin has a phase *lead* over its lateral velocity. This is the force needed to accelerate it against the inertia represented by its own mass and the added mass of fluid.

At the frequencies with which fins of high aspect-ratio are oscillated, these two effects are close to cancelling out. In the simple theory of Lighthill (1970), his equations (70), (61) and (69) with $b = \frac{1}{2}a$ imply that at each instant the sideforce takes its steady-state value multiplied by

$$[F(\sigma)+iG(\sigma)]+\tfrac{1}{2}i\sigma; \qquad (1)$$

here, $F(\sigma)+iG(\sigma)$ with F positive and G negative is the Theodorsen function, representing the Wagner phase lag, and the term $\frac{1}{2}i\sigma$ represents the added mass of water only. The imaginary part of (1), which represents the phase lead or lag, vanishes when the frequency parameter

$$\omega c/U = 2\sigma \qquad (2)$$

(based on radian frequency ω, fin chord c and swimming speed U) takes the value 0·7. Adding a small correction for the mass of the fin itself reduces this figure to 0·6. These are close to the values of 0·7–1·0 observed for scombroid fishes (Fierstine and Walters, 1968).

These simple arguments, which suggest that high-frequency tail oscillations in, say, tuna do not require very special efforts to overcome inertia, are borne out by the more refined analyses of Kambe and Chopra (1976). In the meantime, although the "recoil" analysis of Section 2 allows for the oscillating side-force acting on the caudal fin to have arbitrary phase, the above arguments suggest that we may think of that side-force as being fairly closely in phase with the fin's own lateral velocity.

2. Reactive–Resistive Recoil Theory

Figure 2 indicates the assumptions used in the recoil theory. An oscillatory reactive force of amplitude F and radian frequency ω (so that the period of oscillation is $2\pi/\omega$) acts on the inflexible anterior part of the fish body. This is a mainly inertia-controlled system, so that it responds primarily to the lowest harmonic of this periodic force. We write that as $F\exp(i\omega t)$, on the usual understanding that the *real part* of such a complex expression for an oscillatory quantity is to be taken.

Note that this oscillatory reaction on the body is equal and opposite to the force driving the caudal fin. If that (as just suggested) is in phase with the fin's velocity, then as shown in Fig. 2 the reaction on the body is in antiphase with the fin's velocity. In a *purely* inertia-controlled system, this would make the local displacement of the body in phase with the fin's velocity, as shown. We want to investigate, however, such modifications to this pure reactive response as may result from hydrodynamic resistance.

As a representative distribution of the depth s of the fish body along the length l of its relatively inflexible anterior portion, we take

$$s = s_m\left[4\frac{x}{l}\left(1-\frac{x}{l}\right)\right], \tag{3}$$

as illustrated in Fig. 2. Here the maximum body depth s_m may, of course, be in an arbitrary ratio to the length l.

It is reasonable to take the effective inertia per unit length in response to lateral motions as ρs^2. This includes *both* (i) an added mass of water which, as Lighthill (1970) concludes, can best be taken as $\frac{1}{4}\pi\rho s^2 \doteqdot 0{\cdot}8\rho s^2$, where ρ is the density of water, *and* (ii) the mass per unit length of the fish cross-section which is typically around $0{\cdot}2\rho s^2$.

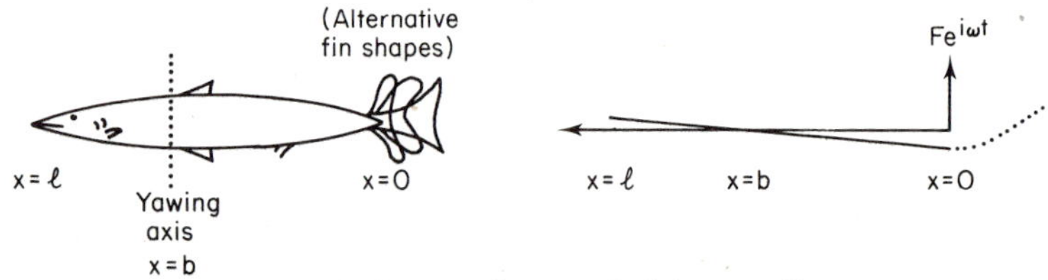

FIG. 2. Assumptions underlying recoil.

I shall make a similarly rough estimate of the *resistance*, per unit length of fish body, to lateral motions with velocity w perpendicular to the backbone. Fish cross-sections of depth s and much smaller width are likely to give almost as much resistance as a flat plate of the same depth. Such a plate moving perpendicular to itself at a steady velocity w experiences a resistance close to $\rho s w^2$ per unit length (Flachsbart, 1935). There is considerable evidence from aeronautics that forces on elongated wing-like surfaces with forms near to flat plates can be estimated quite well by using this formula also in unsteady conditions. I adopt the resistance formula $\rho s w^2$ per unit length as something of an upper limit (corresponding to a drag coefficient of 2 based on area perpendicular to the direction of the velocity) by means of which the maximum possible importance of resistive forces relative to reactive forces can be estimated.

We are interested in how the resistive force may vary in steady swimming when the period of oscillation of the caudal fin is $2\pi/\omega$. We are particularly interested in the response to oscillations of simple-harmonic form

$$w = w_1 \exp(i\omega t), \tag{4}$$

where w_1 as well as w may be complex but, as usual, the real part of w represents the actual lateral velocity of the fish cross-section. This simple-harmonic element is important because higher harmonics are likely to be relatively small in an inertia-controlled response like that of the anterior part of the fish body.

A good approximation to the resistive force acting when w has the form (4) is

$$k\rho s|w_1|w,$$

where

$$k = 8/3\pi = 0{\cdot}85. \tag{5}$$

Figure 3 shows that this differs from the exact resistive force only by higher harmonics (third, fifth, etc.) of relatively small amplitude. In fact, with respect to oscillations at the fundamental frequency, the resistance law behaves effectively as a linear resistance with damping constant $k\rho s|w_1|$. Accordingly, I analyse the system illustrated in Fig. 2 by means of a distribution along the fish body of inertia ρs^2 and damping $k\rho s|w_1|$, each per unit length.

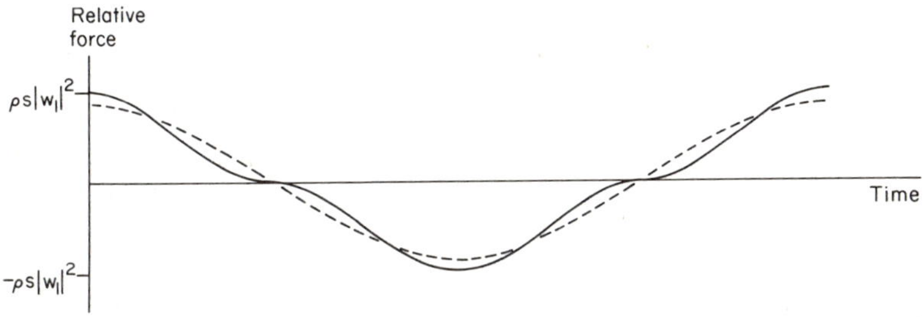

FIG. 3. The plain line shows how the resistive force per unit length, of *magnitude* $\rho s w^2$ and *direction* opposite to w, varies when w oscillates with amplitude w_1. The broken line represents the approximation $k\rho s\,|\,w_1\,|\,w$ to that force (where $k = 8/3\pi$). The error is relatively small and consists entirely of higher harmonics (third, fifth, etc.).

First, I give an analysis of the response which ignores the effect of longitudinal motion through the water at the swimming velocity U. The resulting calculation of how the anterior part of the fish body would react to the oscillatory force at the tail end in the absence of other causes of motion is particularly easy to follow. Furthermore, Lighthill (1970) indicated why, at observed values $U \doteqdot 0{\cdot}1\omega l$ of the swimming velocity, the recoil might be only moderately altered by taking longitudinal motion into account. We later (see Appendix) confirm this in the case here analysed.

In still water, then, the fish of Fig. 2, subjected to the oscillatory force $F\exp(i\omega t)$ at $x = 0$, makes a response in which the lateral velocity has the form (4) and satisfies simple equations of momentum and angular momentum. The momentum equation takes the form

$$i\omega\int_0^l \rho s^2 w_1\,\mathrm{d}x = -\int_0^l k\rho s|w_1|w_1\,\mathrm{d}x + F, \tag{6}$$

where the $\exp(i\omega t)$ factor has been omitted in all terms. The left-hand side represents the rate of change of lateral momentum, integrated up along the length l of the fish. The right-hand side represents the sum of the lateral forces acting, including the resistance (5) integrated along the fish. Similarly, the equation of angular momentum about $x = 0$ is

$$i\omega \int_0^l x\rho s^2 w_1 \, \mathrm{d}x = -\int_0^l xk\rho s |w_1| w_1 \, \mathrm{d}x; \tag{7}$$

this equates the rate of change of angular momentum to the moment of the forces about $x = 0$, so that the force $F\exp(i\omega t)$ *which acts at* $x = 0$ makes no contribution to (7).

Simple experiments (demonstrated in the lecture) indicate that a rigid model of the anterior part of the fish may respond in still water by means of a simple yawing motion

$$w_1 = w_0\left(1 - \frac{x}{b}\right), \tag{8}$$

where $x = b$ represents the yawing axis. Thus, b is a real number, unlike w_0 which is the complex amplitude of the oscillations of lateral velocity at the posterior end $x = 0$ where the force is applied. Actually, equation (8) represents the most general possible distribution of w_1 for a rigid body *except* that the angular velocity w_0/b might not necessarily have the same phase as w_0 (the lateral velocity at $x = 0$).

Numerical evidence, that this simple yawing response represents quite a close approximation to the solution of (6) and (7), is as follows. The left-hand side of equation (7), with w_1 given by (8) and s given by (3), takes the form

$$\frac{4i\omega\rho s_m^2 w_0 l^2}{15}\left(1 - \frac{4}{7}\frac{l}{b}\right), \tag{9}$$

which vanishes at $b = 0{\cdot}571l$. The right-hand side can be calculated as

$$-\frac{k\rho s_m |w_0| w_0}{15 l^2 b^2}[2lb^5 - b^6 - (l-b)^6 + 4l(l-b)^5 - 5l^2(l-b)^4], \tag{10}$$

which vanishes at $b = 0{\cdot}589l$. It follows that choosing the yawing axis in the intermediate position

$$b = 0{\cdot}58l \tag{11}$$

satisfies the angular momentum equation (7) to quite a close approximation whatever the value of w_0. Then w_0 can be derived from the momentum equation (6) which, after the integrals in it have been similarly evaluated, and the value (11) for b has been substituted, becomes

$$0{\cdot}074 i\omega\rho s_m{}^2 l w_0 + 0{\cdot}051\, \rho s_m l |w_0| w_0 = F. \tag{12}$$

Equation (12) with the second term omitted would represent the purely inertial, or "reactive", response

$$w_{\mathrm{R}} = \frac{F}{0{\cdot}074 i\omega\rho s_m{}^2 l}; \tag{13}$$

one in which the lateral velocity $w_{\mathrm{R}} \exp(i\omega t)$ of the point $x = 0$, where the force $F\exp(i\omega t)$ is being applied, responds to it with a phase lag of 90°. Such a pure-imaginary reactive response (13) would correspond to a zero average rate of working by the force.

By contrast, we expect that when the *resistive* second term is included in (12) the complex amplitude w_0 may possess a positive real part, representing a velocity component in phase with the force. That would produce a positive average rate of working by the force $F\exp(i\omega t)$, equal to

$$\tfrac{1}{2}F\,\mathrm{Re}\,(w_0), \tag{14}$$

where Re stands for "real part". This represents additional working by the fish musculature needed to overcome the hydrodynamic resistance to recoil motions; additional, that is, to its working needed to produce the lateral movements of the caudal fin.

To find out just how important in the solution w_0 of equation (12) is this real part, bringing about an energy dissipation rate (14), we rewrite (12) in terms of w_{R}, its pure-imaginary solution (13) in the purely reactive case. Then it becomes

$$w_0\left(1 + 0{\cdot}69\frac{|w_0|}{i\omega s_m}\right) = w_{\mathrm{R}}. \tag{15}$$

Both the real and the imaginary parts of w_0, the solution of (15), are easily computed and are plotted in Fig. 4. Here the abscissa is

$$\zeta = \frac{w_{\mathrm{R}}}{-i\omega s_m} = \frac{F}{0{\cdot}074\omega^2 \rho s_m{}^3 l}; \tag{16}$$

this is the ratio of the reactive response (13) to one in which *lateral displacements* at the posterior end $x = 0$ have an amplitude equal to the maximum body depth s_m. The imaginary part $\mathrm{Im}\,(w_0)$ is always negative, and therefore appears below the axis in Fig. 4, while the real part $\mathrm{Re}\,(w_0)$, generating the average rate of energy dissipation (14), is always positive and appears above the axis.

The general conclusion from Fig. 4 is that whenever ζ is small compared with 1 (for example, when $\zeta < 0{\cdot}2$) the response is almost purely reactive; that is, $\mathrm{Re}(w_0)$ is much smaller than $\mathrm{Im}(w_0)$ and, to close approximation, $w_0 = w_R$. Under those circumstances the rate of energy dissipation (14) is negligible.

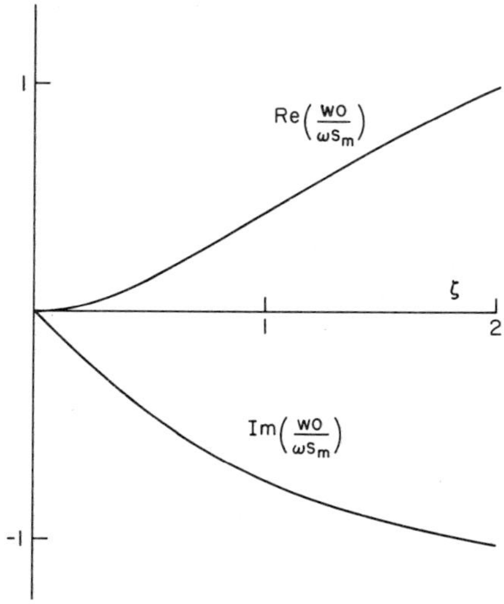

FIG. 4. The real and imaginary parts of $w_0/\omega s_m$ (the complex amplitude of the motions of the posterior end of the nearly rigid part of the body, divided by the amplitude of motions with maximum lateral displacement equal to maximum body depth), are plotted as a function of the parameter ζ defined in equation (16).

The condition $\zeta < 0{\cdot}2$, under which these conclusions hold, means that the recoil motions involve lateral displacements with amplitude less than $0{\cdot}2 s_m$ all along the fish body. To see this, note that (i) the yawing motions specified by (8) and (11) have *greatest* amplitude at the posterior end $x = 0$; while (ii) although the definition (16) of ζ makes ζs_m the amplitude of lateral displacements in a purely reactive response (13), the amplitude of the actual response differs from this negligibly (in fact, by under 3%) when $\zeta < 0{\cdot}2$.

This analysis, then, suggests that a fish whose maximum body depth is large enough for the quantity ζ defined in (16) to be less than 0·2 may have the following advantages: (i) the recoil motions of its body in reponse to the movement of the caudal fin are yawing motions with the greatest amplitude of lateral displacement less than $0{\cdot}2 s_m$ (that greatest amplitude appearing at the posterior end); (ii) furthermore, these motions are primarily a reactive response lagging by 90° in phase behind the forcing effect, and so exacting from it no net rate of working; (iii) in fact, the resistive element in the response (real part of w_0) is at most one-seventh as much, involving lateral displacements of at most $0{\cdot}03s_m$; (iv) accordingly, the extra rate of working by fish musculature to produce these resistive displacements is practically negligible when compared with that involved when the equal and opposite force acting on the caudal fin gives it, typically, much larger displacements.

Although there is of course nothing magic about the number 0·2, it seems useful for indicating a critical value of s_m, particularly since ζ varies rapidly with s_m (by (16), like its inverse cube) and the real part of w_0 varies as the square of ζ. Note, for example, that a 12% reduction in s_m would raise ζ from 0·2 to 0·3; and that the implied 50% increase in the reactive element of the response would be accompanied by a more than doubled resistive element, with lateral displacements of amplitude just over $0{\cdot}06s_m$, giving some quite significant extra dissipation.

The above indications, that there may be a critical value around 0·2 for ζ above which recoil involves significant penalties, were derived under a simplifying assumption. This was the neglect of effects of the fish's longitudinal motion through the water. Lighthill (1970) had indicated that these effects would be relatively small when the ratio $U/\omega l$ takes typical values around 0·1. I briefly verify this in the Appendix, where curves corresponding to Fig. 4 are worked out in that case and found to be very similar (Fig. 5).

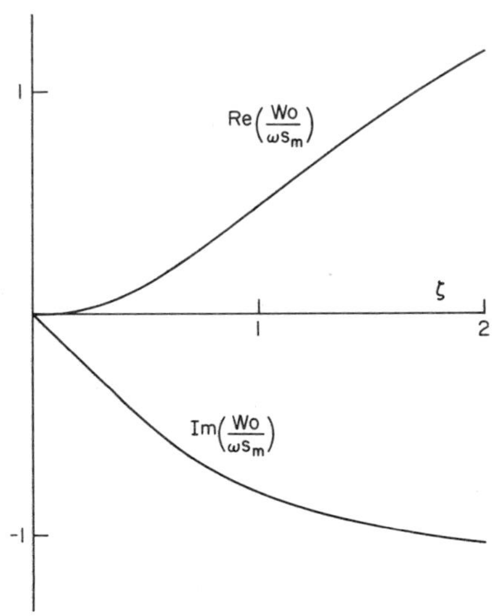

FIG. 5. Calculations for $U/\omega l = 0{\cdot}1$: the real and imaginary parts of $W_0/\omega s_m$ (the complex amplitude of the motions of the posterior end of the nearly rigid part of the body, divided by the amplitude of motions with maximum lateral displacement equal to maximum body depth) are plotted as a function of the parameter ζ defined in equation (16).

The "critical value" of ζ, if we define it as that where the resistive element in the response reaches $0{\cdot}03s_m$, increases from 0·21 to 0·28, but this corresponds to only a 9% reduction in the minimum value of s_m since that depends on $\zeta^{\frac{1}{3}}$.

It may be interesting to investigate in the future how far carangiform motion is in general found in association with maximum body depths s_m large enough for typical values of ζ to remain less than some such critical value between 0·2 and 0·3 (so that F, the maximum side-force on the caudal fin, is less than about $0.02\omega^2 \rho s_m{}^3 l$). This could be interpreted as an adaptation avoiding significant energy dissipation due to recoil motions.

Appendix

In this Appendix, I roughly estimate the errors in the simplified reactive–resistive recoil theory of Section 2 which result from ignoring the effect of longitudinal motion through the water.

Longitudinal motion forces us to distinguish between lateral velocities of two kinds. Following the notation of Lighthill (1970), I use w to mean the lateral velocity of the fish body relative to the water flowing past it and W to mean its absolute lateral velocity. Of the two lateral velocities, it is w which determines hydrodynamic forces (reactive or resistive), but W which determines rate of working. No distinction was required in Section 2 (and in fact w was used) because without longitudinal motion both are the same.

The difference between w and W is the forward velocity U times the angle of yaw. For example, when w is defined by equations (4) and (8), the yawing angular velocity has complex amplitude w_0/b, so the angle of yaw has complex amplitude $w_0/(i\omega b)$, and

$$W = W_1 \exp(i\omega t) \tag{A1}$$

with

$$W_1 = w_1 - \frac{Uw_0}{i\omega b}. \tag{A2}$$

The resistive force opposing the body's lateral motion at velocity $w_1 \exp(i\omega t)$ relative to the water is expected to be exactly as in Section 2, since resistive force depends only on instantaneous relative velocity. The complex amplitude of the resistive force can accordingly be approximated once more as in (5).

By contrast, there is a significant modification to the reactive force, described in Section 2 as due to a total inertia ρs^2 per unit length. Most of this inertia represents the added mass of water; taking the rate of change of the momentum associated with *this* part of the inertia must allow for the fact that the water is moving towards the tail, $x = 0$, at velocity U. The operator

$$\frac{\partial}{\partial t} - U\frac{\partial}{\partial x} = i\omega - U\frac{\partial}{\partial x} \tag{A3}$$

must accordingly be used to obtain the rate of change of that momentum.

I give details for a case when the fish mass and added mass per unit length are, respectively, $\alpha\rho s^2$ and $\beta\rho s^2$. Then the term replacing the left-hand side of (6), which represents rate of change of momentum, becomes

$$\int_0^l \left[i\omega\alpha\rho s^2 W_1 + \left(i\omega - U\frac{\partial}{\partial x}\right)(\beta\rho s^2 w_1)\right] \mathrm{d}x. \tag{A4}$$

This is because the lateral momentum of the fish depends on the absolute lateral velocity W, while the lateral momentum of the water depends on the relative velocity w and changes as described by the operator (A3). Actually, the term proportional to U in (A4) vanishes, because s is zero at both ends of the range of integration (see Fig. 2).

By contrast, the corresponding term in the rate of change of angular momentum about $x = 0$, which is (A4) with an x factor outside the square brackets, can be integrated by parts to give

$$U\int_0^l \beta\rho s^2 w_1 \, \mathrm{d}x. \tag{A5}$$

Remarkably enough, this is almost cancelled by a second term proportional to U in the rate of change of angular momentum: the term obtained when we substitute for W_1 in terms of w_1 by equation (A1). This is

$$-U\int_0^l x\left(\frac{w_0}{b}\alpha\rho s^2\right) \mathrm{d}x; \tag{A6}$$

in fact, equation (7) for rate of change of angular momentum is altered from the case $U = 0$ by the sum of (A5) and (A6), which with s given by (3) is

$$\tfrac{4}{15} U w_0 \rho S_m{}^2 l \left[2\beta - \frac{l}{b}(\alpha + \beta)\right]. \tag{A7}$$

The cancelling is fairly close for $b/l = 0{\cdot}58$; in particular, when the coefficients α and β for fish mass and added mass take the typical values 0·2 and 0·8 mentioned in Section 2, the two terms in square brackets are 1·60 and 1·72. These arguments imply that the choice $b/l = 0{\cdot}58$ continues to satisfy the angular momentum equation (7) to quite a good approximation for moderate values of U such as the values around $0{\cdot}1\omega l$ which are typically observed.

The similarly modified form taken by the momentum equation (6) involves a greater change, largely because there is no cancellation of two terms of opposite sign. Only one term in U is present; it appears when in (A4) we substitute for W_1 terms of w_1 by (A1). It is

$$-\frac{Uw_0}{b}\int_0^l \alpha \rho s^2\,\mathrm{d}x = -0{\cdot}92 Uw_0\,\alpha\rho s_m{}^2, \tag{A8}$$

and must be added to the left-hand side of (12).

Solutions of equation (12) when it has been modified by the additional term (A8) have been calculated for the typical forward velocity $U = 0{\cdot}1\omega l$, and for $\alpha = 0{\cdot}2$, $\beta = 0{\cdot}8$, as a function of the parameter ζ defined in equation (16). Figure 5 plots the corresponding real and imaginary parts of $W_0/\omega s_m$, where the complex amplitude W_0 of the lateral velocity of the posterior end $x = 0$ is

$$W_0 = w_0\left(1+\frac{iU}{\omega b}\right) \tag{A9}$$

by (A1). This quantity W is what determines the rate of working of the force $F\exp(i\omega t)$ as

$$\tfrac{1}{2}F\,\mathrm{Re}(W_0). \tag{A10}$$

As noted in Section 2, the changes from Fig. 4 are found to be only quite modest. This conclusion, that the theory of Section 2 has fairly good accuracy, made me decide to include only that relatively easily intelligible theory in the main text.

References

CHOPRA, M. G. (1974). *J. Fluid Mech.* **64**, 375.
CHOPRA, M. G. (1976). *J. Fluid Mech.* **74**, 161.
DAVIES, D. E. (1963). *Aeron. Res. Council, Rep. and Memor.* no. 3409.
FIERSTINE, H. L. and WALTERS, V. (1968). *Mem. S. Calif. Acad. Sci.* **6**, 1.
FLACHSBART, O. (1935). *Z. angew. Math. Mech.* **15**, 32.
KAMBE, T. and CHOPRA, M. G. (1976). *J. Fluid Mech.* **79**, 49.
LIGHTHILL, M. J. (1970). *J. Fluid Mech.* **44**, 65. (This paper appears also as Chapter 4 of Lighthill, "Mathematical Biofluiddynamics", Society for Industrial and Applied Mathematics, Philadelphia, 1975.)
LOMAX, H. (1962). Indicial Aerodynamics. *In* "AGARD Manual on Aerolasticity" (R. Mazet, ed.), Vol. 2, Chapter 6.
WEIHS, D. (1972). *Proc. Roy. Soc. B*, **182**, 59.

10. Performance and Behaviour on Migration

F. R. HARDEN JONES

Fisheries Laboratory, Lowestoft

1. Introduction

Professor Lighthill has related form to function in mathematical terms; Dr. Wardle has sought to extend the theory by measuring the power developed when fish swim in tanks. I shall try to take the story one step further by telling you something about performance and behaviour in the open sea.

The commercial fisheries are based on migratory fish which winter, spawn and feed in different areas, the distribution of a particular stock often appearing to be contained within a regional hydrographic circulation. There are, for example, stocks of cod at Newfoundland, Labrador, West Greenland, Iceland, Faroe Bank, Faroe, the Barents Sea, the Norway coast, the North Sea and the Baltic. In general terms these sea areas can be identified with more or less clearly defined hydrographic systems. Accepting that there is, at times, significant mixing between the cod at West Greenland and those at Iceland, and that there may be more than one stock of cod in the North Sea, the different stocks are, for the fisheries biologist, independent units which can be described in terms of recruitment, growth and mortality.

The presence of a number of stocks could be thought of as a form of biological insurance to provide protection for the population as a whole in the event of the collapse of one particular stock under adverse conditions. Migration itself is probably an adaptation towards abundance, members of a stock being able to draw upon the resources from a wide area at different stages of their life history and at different seasons. It is perhaps no accident that species of commercial importance are migratory: they are abundant because they are migratory, and they are fished because they are abundant.

2. Migration Circuits

The migration patterns of fish can be summarized in the form of simple triangular diagrams as shown in Fig. 1. Figure 1(*a*) includes the juvenile and adult components of the stock, while Figs. 1(*b*) and (*c*) deal with the mature fish. The mature fish move between their spawning, feeding and wintering areas in a sequence which depends on the season in which they spawn: thus spring spawners would follow the sequence F, W, S (Fig. 1(*b*))

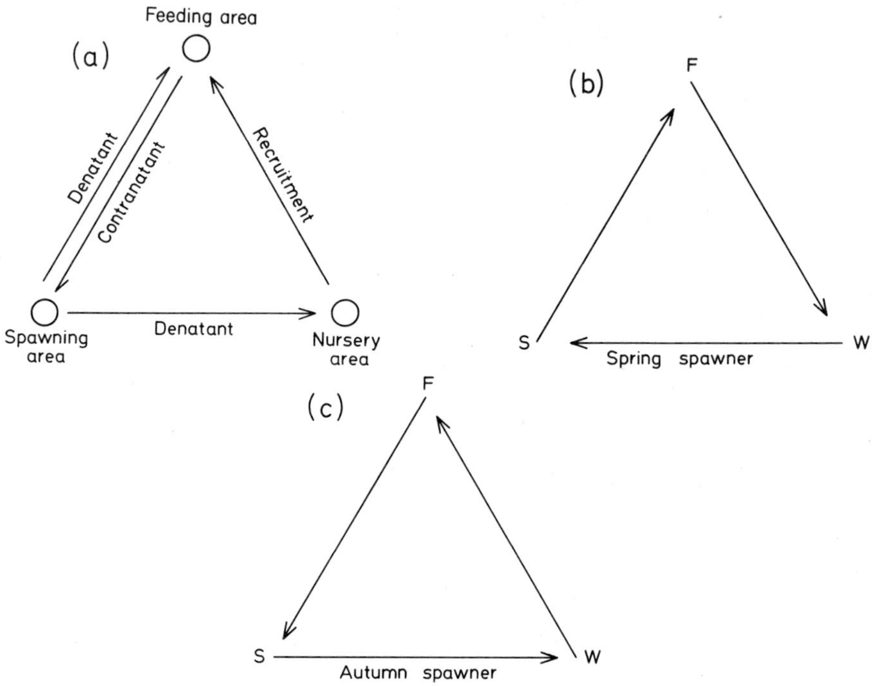

FIG. 1. Diagrams of migration circuits. (a) Juveniles and adults; (b) mature spring spawners; (c) mature autumn spawners. F. S and W; feeding, spawning and wintering areas, respectively.

and autumn spawners the sequence F, S. W (Fig. 1(*c*)). The distance covered during year—the length of the migration circuit—varies between stocks. For example, as shown in Table 1, the migration circuits of cod and herring

TABLE 1. Distance, in km, covered each year by herring and cod when completing their migration circuit

	Herring	Cod
Length of fish	30 cm	80 cm
North Sea	1600 km	1300 km
Northern waters	3000 km	2600 km

in the North Atlantic are about twice as long as those of cod and herring in the North Sea. But in both areas the fish complete the circuit in 1 year. It is true that the Norwegian herring are somewhat bigger than North Sea herring, that there may be bigger cod in the Barents Sea than ever came out of the North Sea, nevertheless, within a species, many fish of the same size appear to complete the longer as readily as the shorter migration, and between species a 35 cm herring may cover the same distance during a year as a 90 cm cod.

Is there any relation between the size of the fish and the distance covered during migration? It is difficult to give a very reliable answer to this question. Conventional tagging experiments—catching, tagging and releasing a fish here, and later recapturing the same fish there—are not very satisfactory. One does not know how long the fish had been waiting to be caught at the recapture position: the estimate of speed can only be a minimum speed.

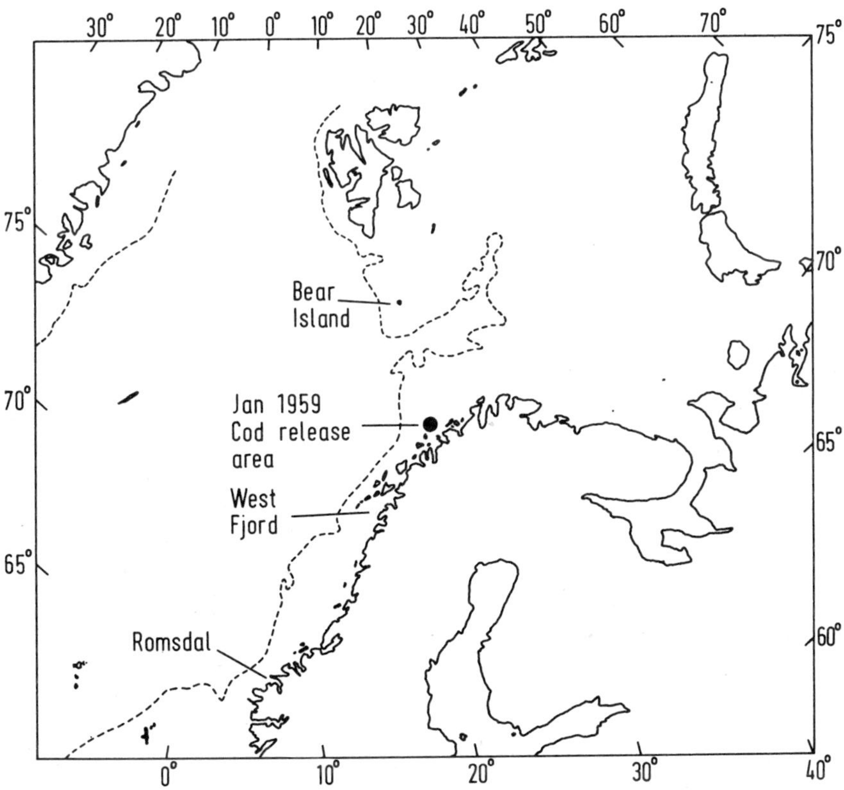

FIG. 2. Arcto-Norwegian cod. Place names of interest in connection with G. C. Trout's tagging experiment at Fugloy Bank, January 1959. The 400 m contour is shown.

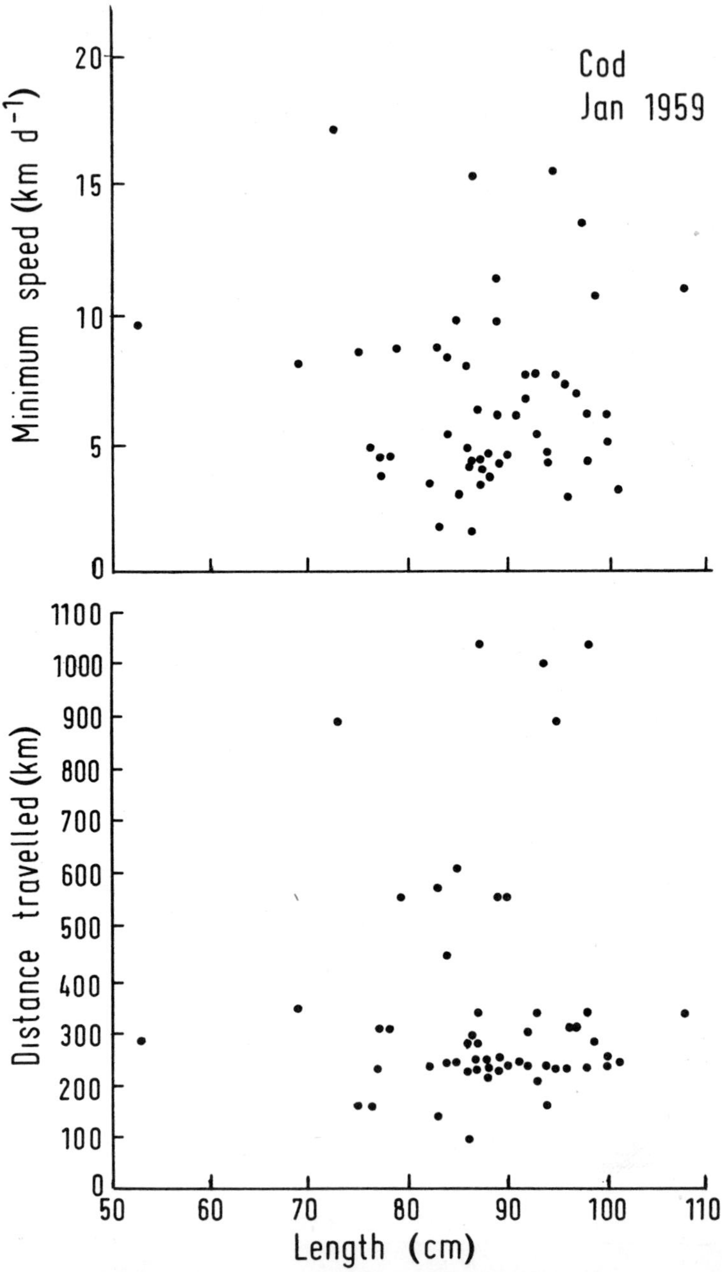

FIG. 3. Arcto-Norwegian cod. The relation between fish length, distance travelled and minimum speed as given by G. C. Trout's tagging experiment.

The best that can be done is to examine the distribution of recoveries within a relatively short time of release. I have analysed the recoveries from two tagging experiments, one with cod and the other with plaice, carried out by my colleagues, Mr. G. C. Trout and Dr. R. C. A. Bannister, respectively. I am grateful to them for letting me use their unpublished results.

Consider first the results with cod. The cod of the Barents Sea winter to the west and south of Bear Island before moving on to their spawning area which is centred on the West Fjord, within the Lofoten Islands. The mature fish leave Bear Island in December, arrive at the Norwegian coast in January

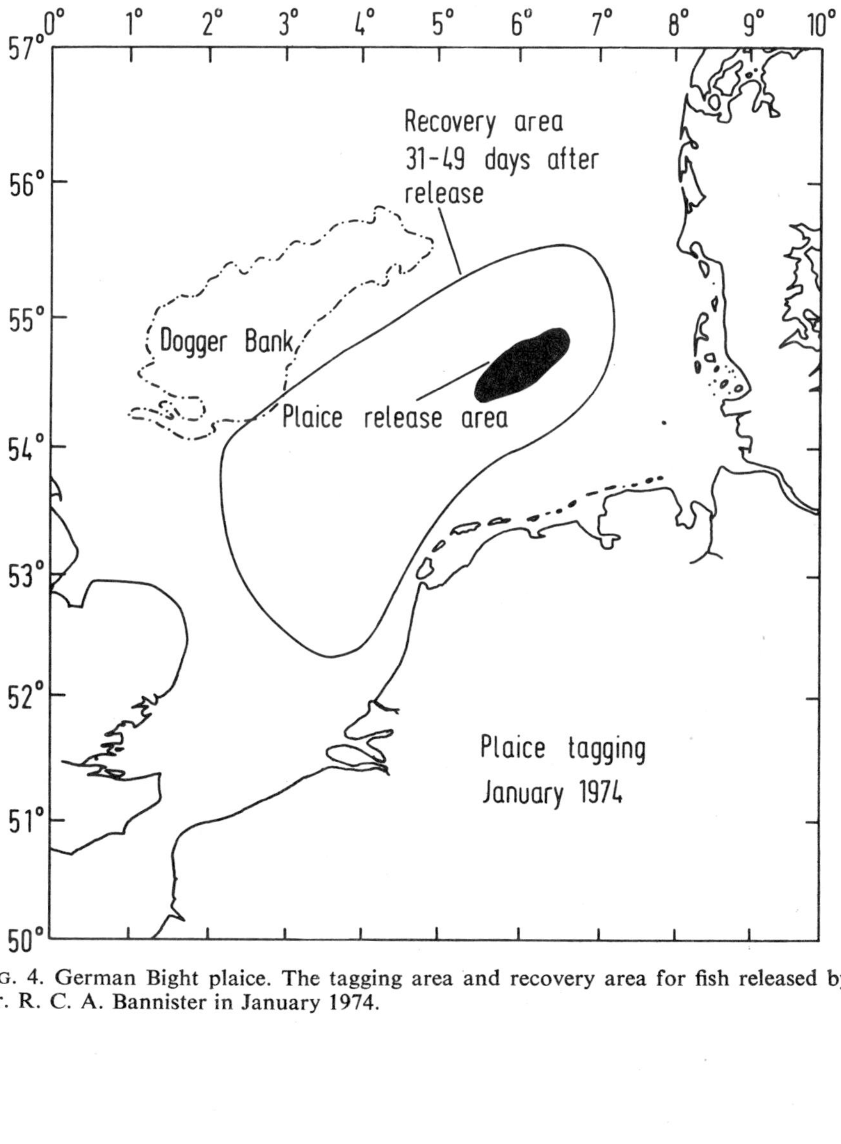

FIG. 4. German Bight plaice. The tagging area and recovery area for fish released by Dr. R. C. A. Bannister in January 1974.

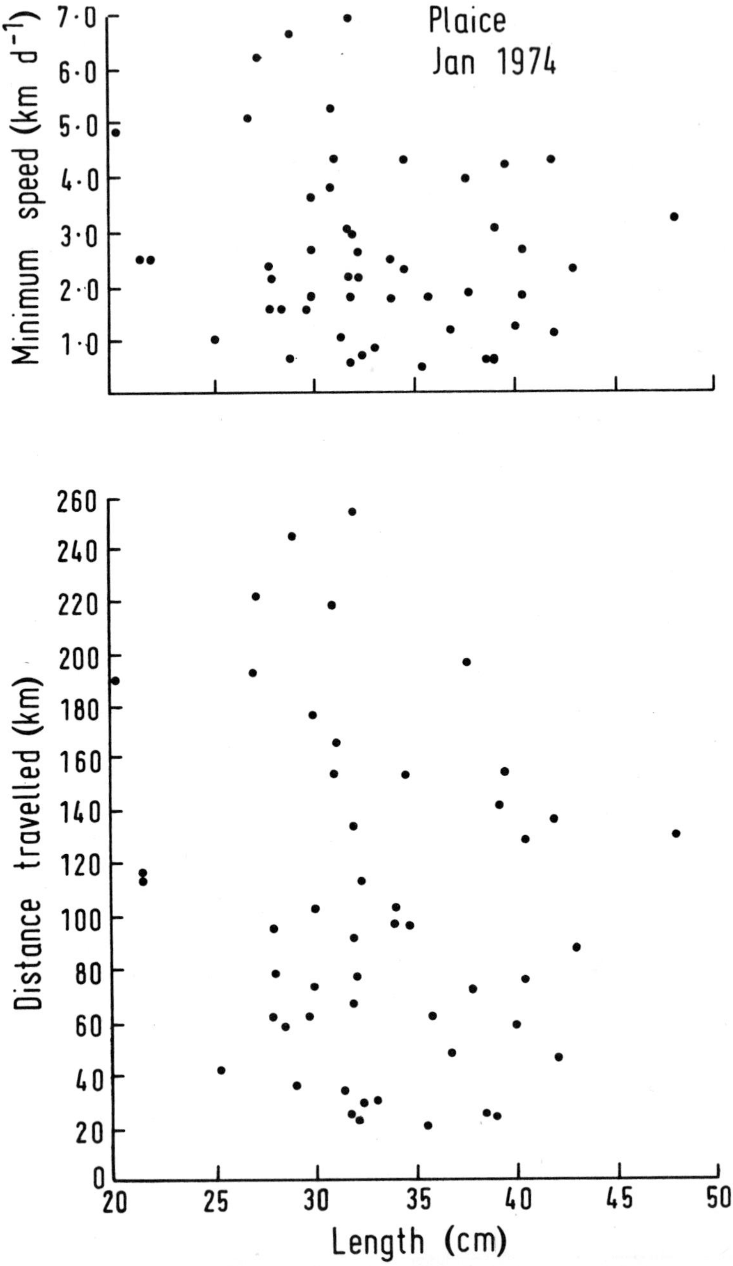

FIG. 5. German Bight plaice. The relation between fish length, distance travelled and minimum speed as given by Dr. R. C. A. Bannister's tagging experiment.

and spawn at Lofoten in March–April. In January 1959, cod were caught, tagged and released at Fugloy Bank. Fifty-four of these fish were subsequently recaptured during the next 90 days. The position of the recaptures and the relation between fish length and distance travelled (in km), speed (in km day^{-1}) and fish lengths per second ($L\ s^{-1}$) are shown in Figs. 2 and 3. There is no obvious relation between fish length and distance travelled, or between fish length and speed.

There are three stocks of plaice in the North Sea and one of these stocks spawns in the German Bight in January–March. In January 1974 Dr. Bannister tagged and released several hundred plaice in the spawning area. Data relating to the position and speed of movement of plaice recovered 31–39 days after release are summarized in Figs. 4 and 5. Here again there is no relation between fish length and distance travelled and fish length and speeds.

The speeds at which cod and plaice travel are rather slow and this is generally true for fish on migration, as can be seen in Table 2. Even the large bluefin tuna and albacore, which are known to make transoceanic migrations in 60 days, are not travelling very quickly. If the Reynolds numbers (R) are calculated from the speeds (V) and lengths (L) in Table 2 using the relation

TABLE 2. Minimum speeds of fish on migration estimated from tagging experiments

Species	Mean length (cm)	Speed			Author
		km d^{-1}	cm s^{-1}	$L\ s^{-1}$	
Sole	30	7–16	9–18	0·28–0·53	Anon. (1965)
Plaice	35	1–7	2–8	0·06–0·23	Banninster (unpubl.)
Herring	25	4–30	5–35	0·20–1·40	Bolster (1955)
Mackerel	35	16–23	19–27	0·54–0·77	Bolster (1974)
Salmon (Sockeye)	70	9–22	10–25	0·15–0·36	Harden Jones (1968)
Cod	80	6–28	7–32	0·09–0·40	Trout (unpubl.)
Albacore	77	26–44	30–51	0·39–0·66	Clemens (1961)
Bluefin	250	93–185	108–214	0·43–0·86	Mather, et al (1977)

$R_L = 100VL$, the values obtained are consistent with low drag coefficients; the fish have no overwhelming hydrodynamic problems. While this may well be true, it is doubtful if such calculations are justified. The speeds given in

Table 2 are over the ground, while the appropriate speed for the calculation of the Reynolds number is through the water.

3. The Two Speeds

A distinction can be made between the two speeds: it could be important to travel quickly over the ground if time were short, or if there were advantages to be gained by arriving before the others. If the journey was made quickly the maintenance energy expended when travelling could be saved and for a small fish, a quick journey might reduce the risk of being eaten on the way. If unable to travel quickly, the risk of death by predation might be reduced by travelling in company; the fish convoy is the fish shoal (Brock and Riffenburgh, 1960). Small fish should go together; but big fish can go it alone.

But if there is something to be said for travelling fast over the ground, there is also something to be said for going slowly through the water. It is this: the power to be exerted against total drag is proportional to the square of the fish's velocity relative to the water; an energy-conscious fish would therefore swim slowly. Weihs (1973) has dealt with this line of argument in detail.

So it could be argued that when on migration the wise fish would make arrangements to move quickly over the ground while moving slowly through the water. I hope to show you that the paradox is, perhaps, more apparent than real.

We know little or nothing about the manner in which fish move when on migration. But we are starting to learn something about how fish move from one position to another by using a variety of tracking techniques. Interest in this sort of work is world wide and is increasing. At Lowestoft we have been studying the movements of fish in the open sea. I shall discuss the work in progress and the results obtained with plaice and cod. In some respects the work is incomplete, in others not yet fully understood; some aspects of the work, however, are now in their final stages of preparation for publication.

4. Tracking Work with Plaice

At Lowestoft we have combined two acoustic techniques into a reliable and effective tracking system which can be used from a research vessel at sea. In the late 1960s we were fortunate to have the loan of the Admiralty Research Laboratory's 300 kHz sector-scanning sonar (Cushing and Harden Jones, 1967) which was installed in our research vessel *Clione*. The equipment has been fully described elsewhere (Voglis and Cook, 1966; Mitson and Cook,

1971). The scanning sonar has a high resolution, 0·3° in bearing and 7 cm in range, and, as fitted on R.V. *Clione*, can scan horizontally (to give bearing in azimuth) or vertically (to give depth). When used in conjunction with the transponding acoustic tag invented by Mitson and Storeton-West (1971) it has been possible to follow individual fish in the open sea (Greer Walker, Mitson and Storeton-West, 1971). Thus plaice have been followed in East Anglian waters for periods up to 55 hours and for distances up to 60 km. In these experiments a tagged fish was released and followed by the research ship, and kept under continual surveillance at ranges between 100 and 250 m. During the tracking the ship was positioned regularly at 15 minute

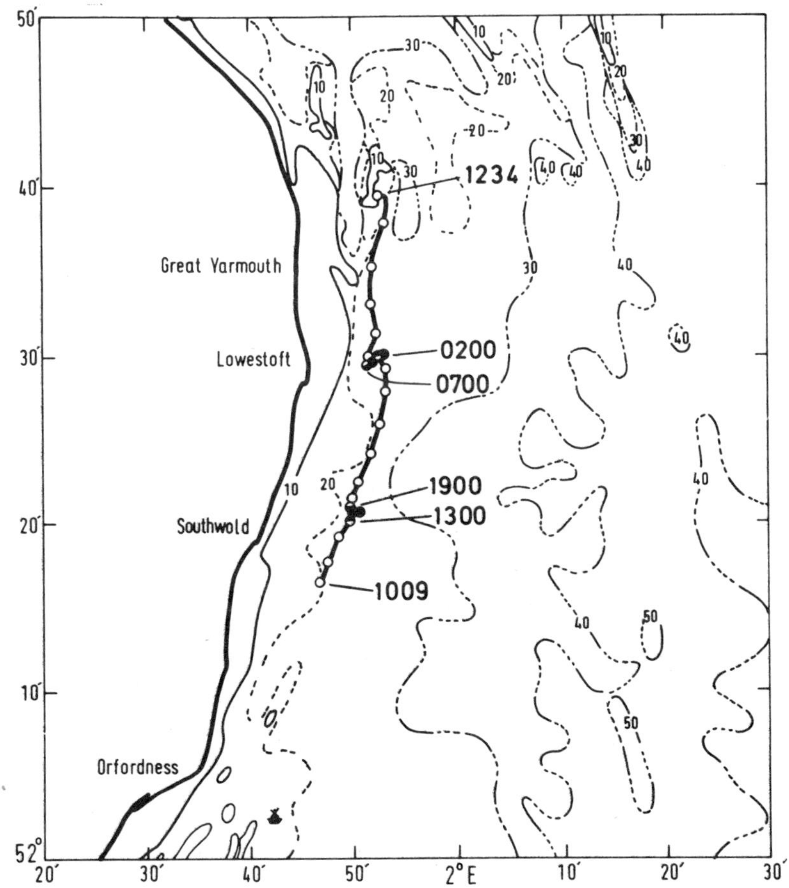

FIG. 6. Track chart of Plaice 7, 41·0 cm long, released at 1009 hours GMT, 12 December 1971. Hourly positions of the fish are indicated and the times of slack water are given. ○, North-going tide; ●, south-going tide; ◒, low-water slack; ◓, high-water slack.

intervals by reference to the Decca Navigator System which was accurate to ±50 m in the working area. The signal returned by the transponding tag was clearly recognizable on the B-scan of the sonar display and allowed the fish to be positioned relative to the ship to within ±2·5 m in range, 1·0° in bearing and 1–2 m in depth. Twelve tracks have been made with plaice off the Suffolk coast (Greer Walker, Harden Jones and Arnold, 1977). Here I I shall describe the tracks on one of these fish.

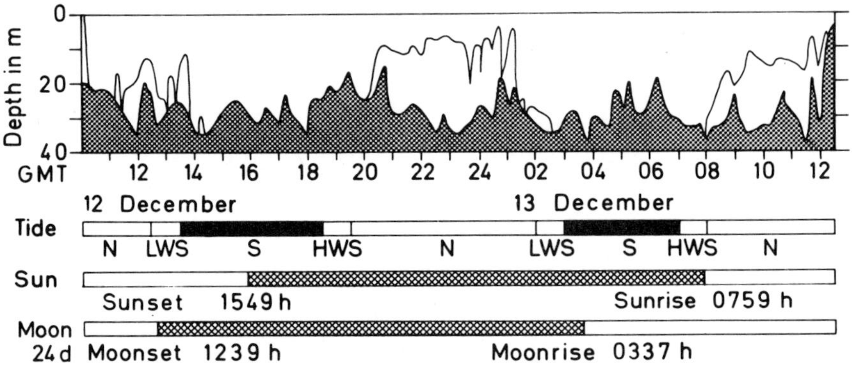

FIG. 7. The depth of Plaice 7 in relation to the direction of the tide and other environmental factors.

TABLE 3. Speed over the ground, direction of tide and position in the water column. Data for Fish 7 rounded off and simplified from Fig. 6

Time (GMT)	Speed over ground (cm s^{-1})	Direction of tide	Position in water column
1000–1300	80	North	Midwater
1300–1900	9	South	On bottom
1900–0200	68	North	Midwater
0200–0700	8	South	On bottom
0700–1230	105	North	Midwater

Plaice 7, released on 12 December 1971, was tracked for 26 hours during which it moved 43 km to the north. The track chart (Fig. 6) shows that the fish gained ground to the north during the northerly tides, the fish coming into midwater after high-water slack and returning to the bottom at low-water slack (Fig. 7). These vertical movements, which occur about every 12 hours, can be described as semi-diurnal, and are related to the tidal cycle rather than to that of the sun. The relation between tide, the position of the fish in the water column and distance moved over the ground are summarized in Table 3. The fish was abandoned at 1234 hours, on 13 December when it "went aground" in shallow water at the Cross Sand. Over the period of the tract its speed was 39·4 km day^{-1} or 45·6 cm s^{-1} equivalent to 1·11L s^{-1}. It is clear that the speed of the fish over the ground was greatest when the tide was set in the direction of overall movement.

The semi-diurnal pattern of vertical migration locked to the tidal cycle was, with the exception of one plaice, characteristic of all those that moved a point-to-point distance of 15 km or more during the period of surveillance. The ground speeds of the tracked plaice ranged from 16 to 40 km day^{-1} (18 to 46 cm s^{-1}) or in terms of lengths per second, 0·5–1·0L s^{-1}. These figures are comparable with those summarized in Table 2. But what about speed through the water? Measurements of the water current were made during some of the tracks; from a second research vessel (R.V. *Corella*) during the track of fish 7 and from moored current meters during other tracks. Comparisons could therefore be made between the ground velocity of the fish, as determined from fixes made every 15 minutes, and the water current velocities made within a few kilometres of the fish. We have not yet completed our analyses of these results but comparisons made show that when the fish are in midwater, their velocity over the ground is usually similar to that expected if they were being carried passively by the tide. There were exceptions. But when there was a relative velocity between fish and water it was low, and rarely exceeded 40 cm s^{-1} or 1L s^{-1}.

So the pattern of behaviour often observed during these tracking experiments does allow the plaice to go quickly over the ground while moving slowly through the water. We think that the behaviour can be adequately described by the phrase "selective tidal transport" and Dr. Weihs (1977) has considered the advantages of this method of migration in terms of energy cost. It is obvious that selective tidal transport raises interesting sensory problems in relation to leaving and returning to the bottom but these matters which deal with the "selection" side of the fish's behaviour are beyond the scope of this paper.

5. The Tidal Transport Hypothesis

The tracking results have led us to suggest that plaice select the appropriate tide for transport when on migration and we have tried to apply this hypothesis to the stock which spawns in the Southern Bight of the North Sea with peak egg production in January. The spawning area is indicated in Fig. 8, the mature plaice moving from south of the Dogger Bank into the Southern Bight and so towards the spawning grounds in November and

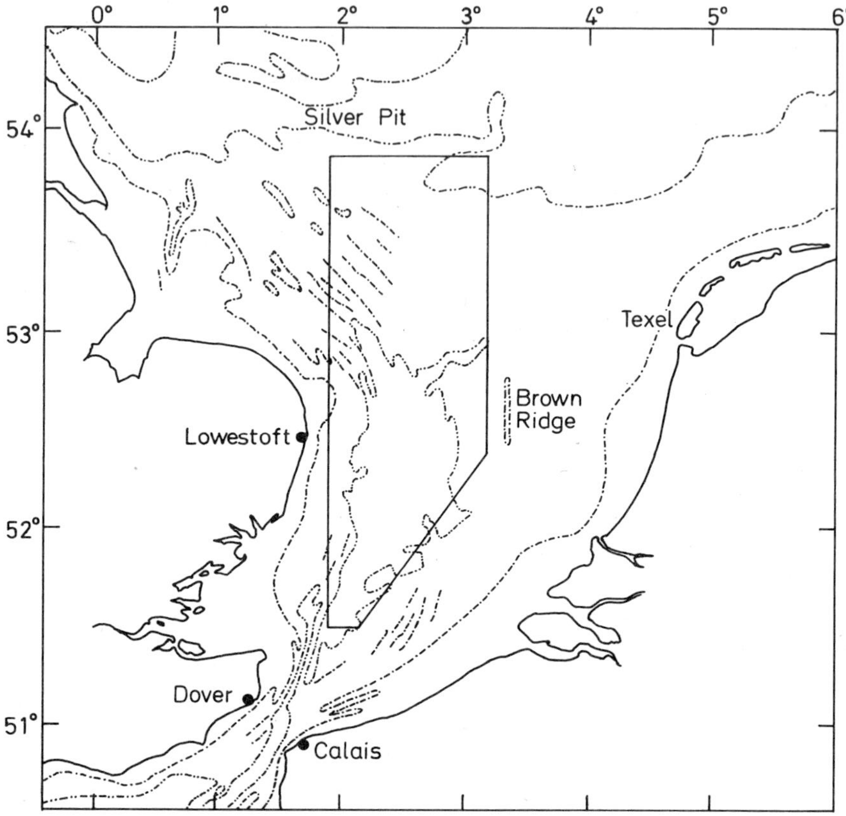

FIG. 8. A chart of the Southern Bight of the North Sea. The plaice spawning area lies between 51°30′–52°00′ N and 2°–3° E. In summer many of the Southern Bight spawners feed to the north of latitude 53° N. Midwater trawling to test the selective tidal transport hypothesis was carried out within the area indicated. The 10 fathom and 20 fathom depth contours are shown.

December. According to the selective tidal transport hypothesis, mature plaice on their spawning migration should come off the bottom into midwater on the south-going tides and return to the bottom on the north-going tides. We have tried to test the hypothesis by fishing along the line of the migration route with midwater trawls and comparing the catch of plaice taken on tows made on northerly and southerly tides. If the plaice migrate in this way, we would expect the catch of a southerly tow to exceed that of a northerly tow.

To test the hypothesis we have adopted a paired haul technique, a research vessel towing an Engel midwater trawl for 3 hours during the peak of one tide, followed, 3 hours later, by a second 3 hour haul during the peak of the next tide. Both hauls were made over the same ground. Each haul was made in the direction in which the tide was running and each pair allowed a comparison to be made between the catch of plaice in midwater on northerly or southerly tides. The headline of the net was set so that the gear fished that part of the water column which the tracking exercises showed plaice to occupy when they were in midwater. The work is being spread over three seasons, of which two have been completed. Our results to date are summarized in Table 4 and they appear to confirm the tidal transport

TABLE 4. A comparison between midwater tows for plaice made on consecutive northerly and southerly tides in the Southern Bight of the North Sea November–December 1974–5 (Harden Jones, Greer Walker and Arnold, in preparation)

Number of valid pairs of tows	Southerly tide tow greatest	Northerly tide tow greatest	Tows equal
41	33	6	2

hypothesis. In terms of numbers of fish, 237 plaice were caught when the tide was running south and 66 plaice were caught when the tide was running north. Moving a step further, we now suggest that after spawning the spent fish return to the northern feeding area by a similar mechanism. Thus we might expect, in February and March, that the tow on a northerly tide should catch more spent fish than the tow on a southerly tide. We are looking into this matter. (Recent results obtained by Dr. Arnold have confirmed this prediction (Report R.V. *Clione* Cruise 2, 31 January–6 February 1976, Fisheries Laboratory, Lowestoft).

We have not finished with the plaice tracks, and the more we look at them the more we see in them. For example, the ground speed of one midwater plaice showed regular variations about the mean water velocity as measured at a nearby current meter. Do these changes reflect the accelerations and decelerations of the tidal current which could have been produced by a series of regular and substantial sandwaves which ran normal to the tidal axis and thus to the fish's line of movement? The height to which the plaice rise when in midwater is another puzzling feature of the behaviour pattern. In many cases a fish changed depth so as to keep a more or less constant distance above the bottom; it was almost as if the fish was following a flow line.

A plaice has a density within the range of 1·063–1·066 (Harden Jones and Marshall, 1953) and will sink in sea water of density 1·026. The plaice, which has no special buoyancy organ, must exert a force equal to about 5% of its body weight to stay in midwater. When the drag forces associated with the lift are taken into account, the energy cost to the fish will be about 7% of that required when swimming slowly (Alexander, 1967). We have wondered if the plaice is obtaining any help in reducing the energy cost of lift by making use of forces generated by the flow of water over a wavy bottom. (In discussion Professor Lighthill commented that he thought it unlikely that plaice would take advantage of lift forces generated by flow over the sea bed. With a reasonable angle of attack, a plaice would only need to swim relative to the water at a speed of 10–20 cm s^{-1} to obtain the lift necessary to stay in midwater.) Lift forces must be present as Woodhead sea bed drifters, fitted with acoustic transponding tags and tracked in the same area, rise off the bottom into midwater at tidal flows over 50 cm s^{-1} (Harden Jones, Greer Walker and Arnold, 1973; Dickson, 1976).

The lateral compression of the plaice and other flatfish is usually regarded as an adaptation to a demersal life on the bottom. I suspect that this is not the whole story. If our ideas are correct, migrating plaice may spend half their time in midwater. Could their shape also be regarded as an adaptation towards gliding or sail-planing? The suggestion that the plaice could be regarded, in part at least, as a pelagic rather than a demersal species also provides an alternative explanation for the lack of pigmentation on their left hand (or lower) side. ("Fishes habitually swimming at or near the surface . . . —are coloured silvery or white on the belly and sides . . ." (Norman, 1963, p. 177).)

6. Buoyancy Problems in Cod

Dr. Greer Walker is interested in tracking cod; do they behave in the same way as plaice? The cod, and other gadoids, have resolved the buoyancy problem by using a gas-filled sac, the swimbladder, which adds to its volume without significantly increasing its weight. We have found that cod, adapted to neutral buoyancy in shallow aquarium tanks, have a swimbladder whose volume is equal to $5{\cdot}01 \pm 0{\cdot}33\%$ on a volume/weight basis. Rounding this off to a convenient 5% allows one to construct the Ideal Cod as detailed in Table 5.

But solving the buoyancy problem with a gas-filled swimbladder has raised other problems in the cod and related fish. The swimbladder is closed and gas can only be added by secretion and removed by resorption: the swimbladder has two special areas—the secreting gas gland and the resorbing oval—to deal with these physiological processes. Within reasonable limits, the volume occupied by the swimbladder gas can be assumed to obey Boyle's

TABLE 5. The composition of an Ideal Cod in neutral buoyancy with sea water of density 1·026

Weight	1 kg
Density of tissues	1·081
Volume of tissues	925 ml
Volume of swimbladder	50 ml
Total volume of fish	975 ml

Law. When the fish swims up in the water column, the pressure decreases; the volume of the swimbladder and the fish as a whole increases; and its density becomes less. The reverse will take place when the fish moves down in the water column. Under these circumstances the fish must maintain its position at the new depth by active compensatory swimming movements until the swimbladder volume has been adjusted by the addition or removal of gas. This takes time and it is important in the cod tracking work to know

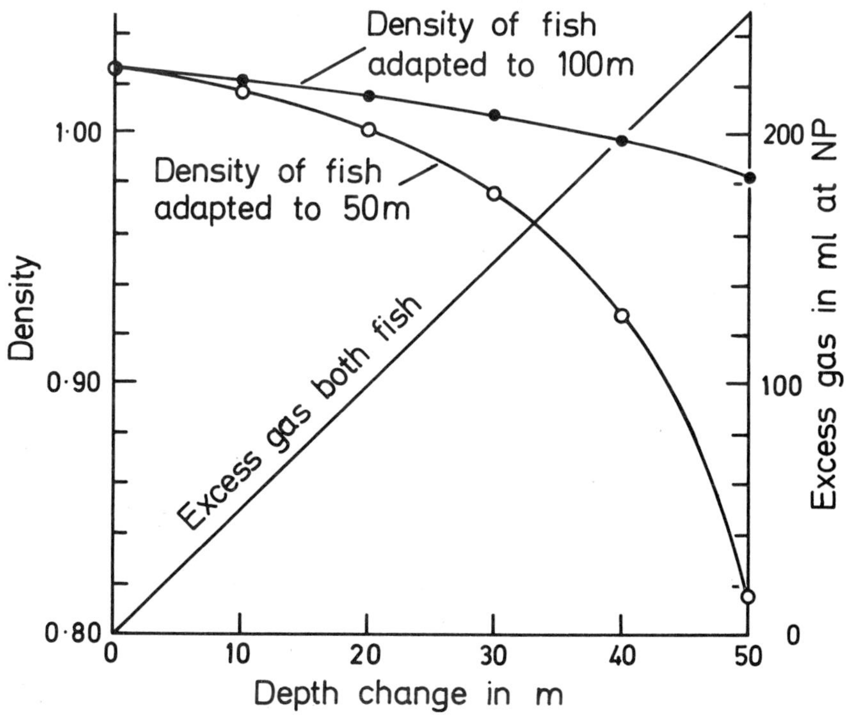

FIG. 9. The buoyancy problems faced by cod when moving from 100 to 50 m and from 50 m to the surface. Data from Table 6.

how much time so that arrangements can be made to allow a fish to adjust the swimbladder volume to the depth of water at which it is swimming.

A pressure change of one-tenth of an atmosphere, which will be incurred when the fish moves up or down in the water column by 1 m, will increase or decrease the swimbladder volume by one-tenth when the volumes are referred to normal pressure. Thus the volume of the Ideal Cod will change by 5 ml for every metre; more generally a vertical movement of 1 m corresponds to a change in swimbladder volume of 5 ml kg^{-1}. For example, the buoyancy problems to be faced by cod when moving up the water column are summarized in Table 6 and the data are plotted graphically in Fig. 9. While the volume of gas at normal pressure which has to be removed or added to the swimbladder depends directly on the extent of the vertical excursion, the density change of the fish, and thus the energy to be expended on maintaining its position in the water, depends on the proportional pressure change. Thus the gas problem is the same for fish moving through a similar depth change in deep or shallow water. But fish living in deeper water have an advantage so far as density changes are concerned and in this connection the column in Table 6 headed "Sinking factor" (the term "sinking factor" was introduced by Lowndes (1937) and is numerically equal to a 1000 times the quotient obtained by dividing the density of the fish by that of the water in which it is living) is some measure of the work to be done against positive buoyancy.

Dr. Greer Walker decided to keep his cod, already fitted with acoustic transponding tags, in cages on the sea bed and to release them after a suitable period of adjustment. He wanted to know how long the cod should be left in the cages. The experimental problem concerns the rates of gas secretion and resorption and I shall describe, briefly, what we have done. Full details of this work will be published in due course (Harden Jones and Scholes, in preparation).

The experimental apparatus (Fig. 10) was a cylindrical steel tank with perspex viewing ports at either end. The fish was put in through a circular hatch on the top of the tank. Water pressure was varied by adjusting the inlet and outlet valves against a 20 m or 30 m head, depending on whether we were working at Lowestoft or Plymouth. Changes in the pressure to which the fish was adapted were determined by noting the pressure head at which the fish was in hydrostatic equilibrium. With a cooperative fish these observations only took a few minutes and were repeatable to within $\pm 0{\cdot}3$ m.

Working with 350–650 g cod we have found that the rate of gas secretion is slow, within the limits of our experiments is independent of pressure and

TABLE 6. Changes in volume and density at intermediate and final depths when an Ideal Cod (weight 1 kg, volume 975 ml, swimbladder volume 50 ml) moves up from 50 m to the surface and from 100 m to 50 m

Depth (m)	Pressure (atmospheres)	Proportional pressure change as % of initial pressure	Volumes (ml)			Density	Sinking factor
			Swimbladder	Excess gas at normal pressure	Total for fish		
50	6	—	50	0	975	1·026	1000
40	5	16·6	60	50	985	1·015	989
30	4	33·3	75	100	1000	1·000	975
20	3	50·0	100	150	1025	0·975	950
10	2	66·6	150	200	1075	0·930	906
0	1	83·3	300	250	1225	0·816	795
100	11	—	50	0	975	1·026	1000
90	10	9·0	55	50	980	1·020	994
80	9	18·0	61	100	986	1·014	988
70	8	27·3	69	150	994	1·006	980
60	7	36·4	78·5	200	1004	0·999	974
50	6	45·4	91·6	250	1017	0·984	959

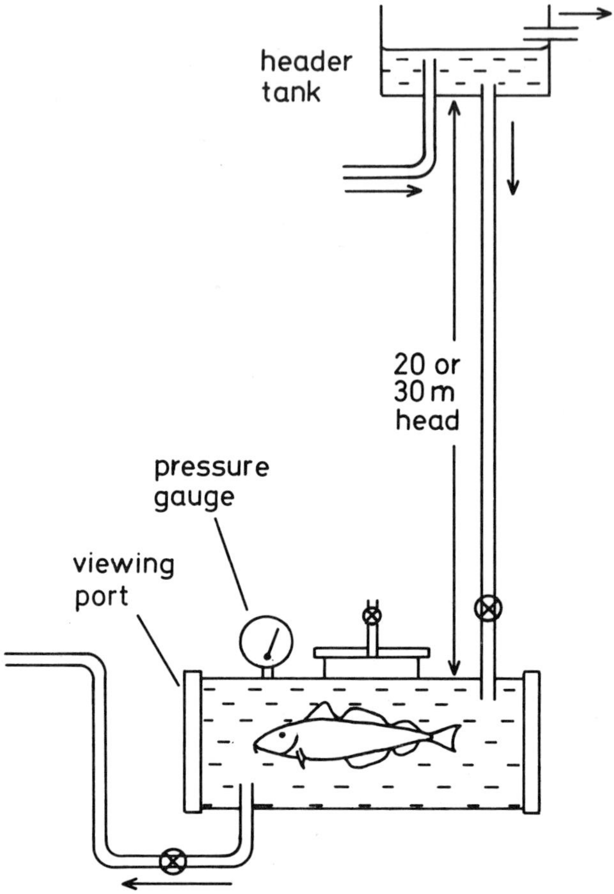

FIG. 10. Apparatus to determine the rates of gas secretion or gas resorption in cod subjected to an increase or a decrease in hydrostatic pressure. Pressure changes within the tank were controlled by adjusting the inflow and outflow valves.

is greatly reduced at low temperatures (Table 7). By contrast the rate of gas resorption is relatively fast, is pressure dependent (Fig. 11) and is hardly affected by temperature.

The difference between the rate of gas resorption and that of gas secretion is substantial in shallow water and becomes very much greater in deeper water. This must pose some problems to the cod with regards to diurnal vertical migrations and can be illustrated by a simple model. Let us suppose that a cod leaves the bottom at sunset and swims towards the surface at a rate which allows it to maintain neutral buoyancy during the migration,

TABLE 7. Rate of gas secretion in the cod and equivalent rates of descent in relation to temperature. Data from Harden Jones and Scholes (in preparation)

Temperature (°C)	Secretion (ml kg^{-1} $minute^{-1}$)	Equivalent rate of descent (m $hour^{-1}$)
0	0·018	0·216
5	0·048	0·576
10	0·077	0·924
15	0·107	1·284

excess swimbladder gas being removed by resorption. After 1 hour the migration stops, and the cod stays at the new depth until it is time to return to the bottom at a rate which allows it to maintain neutral buoyancy and to reach the bottom at sunrise. Our experimental data on gas resorption can be used to calculate the height to which the fish could rise above the sea bed in

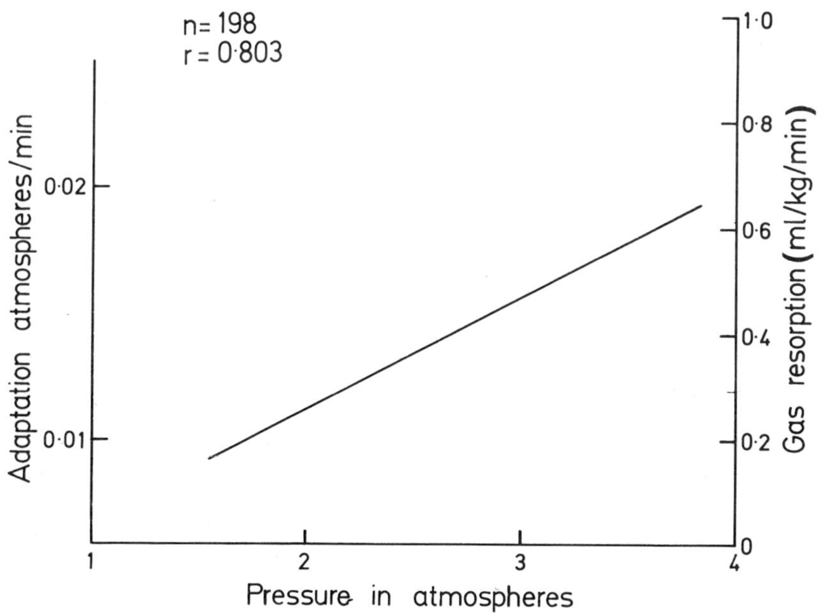

FIG. 11. Gas resorption in the cod in relation to pressure. The rates of adaptation, in terms of $\delta P/T$, increases with pressure, the slope term in the equation $y = kx+c$ being $k = 0{\cdot}0043$, where y = pressure change (in atmospheres) to which the fish adapts per minute and x = pressure (in atmospheres) at which the fish is in neutral buoyancy. The equivalent rates of gas resorption in ml kg^{-1} $minute^{-1}$ are also given.

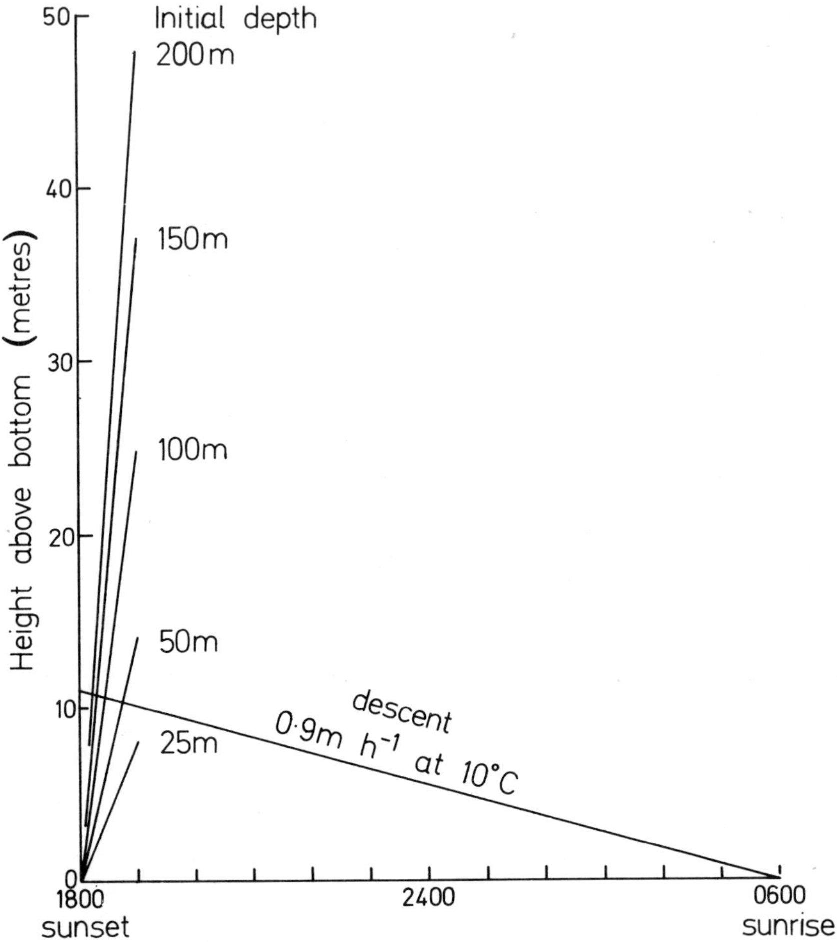

FIG. 12. A diagram to show the limitations that the swimbladder could impose to the diurnal vertical migrations of a cod. The height to which a cod could swim off the bottom in 1 hour was calculated from the relation $P_1 = P_0 \exp(-kT)$, where P_0 is the initial pressure, P_1 the final pressure, T time in minutes (here 60 minutes) and $k = 0{\cdot}0043$ (see Fig. 11).

1 hour when initially adapted to different depths. As shown in Fig. 12, the fish adapted to deeper water would be expected to come further up the water column and at a faster speed of ascent, the rate of gas resorption being proportional to pressure. But whatever depth they come from, all the cod must return along the same descent line, 0·9 m hour^{-1} at 10 °C, if they are to reach the bottom in neutral buoyancy by sunrise.

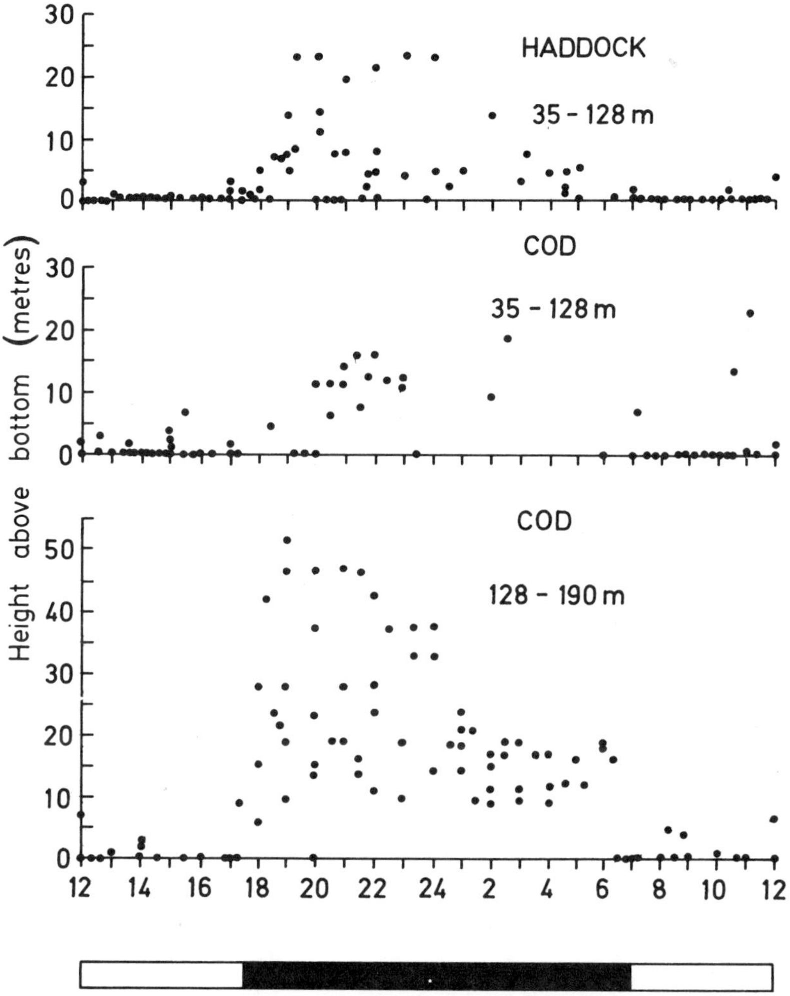

FIG. 13. The extent of diurnal vertical migrations of cod and haddock in Canadian waters as estimated by Beamish (1966, Fig. 4) from echo sounder records.

There are few quantitative data on the extent of the diurnal vertical migrations of cod and other gadoids. Echo sounder records obtained by Ellis (1956) at Skolpen Bank (Barents Sea) showed little difference in the average depth of cod by day and by night. While the shoals were more dispersed during the night, vertical movement off the bottom did not exceed 55 m in depths of 210 m. Beamish (1966, Fig. 4) has analysed echo sounder records taken in Canadian waters to show the vertical distribution of echoes, thought to be returned by cod and haddock. His data, here redrawn as Fig. 13, suggest that cod from 120 to 190 m rise to a greater height above the bottom, and do so more quickly, than cod and haddock from 35 to 128 m. While the distribution of the echoes from the deep-water cod show a steady return towards the bottom, the data do not give any information as to the movements of individual fish. Further evidence of extensive vertical movement by cod in the Gulf of St. Lawrence is given by Brunel (1972), but here again, as with most echo sounder data, we cannot say what the individual fish are doing.

While the extent of the vertical movements detected by echo sounding is more or less in line with that suggested by the experimental model, Beamish's results suggest that some fish are staying well off the bottom for

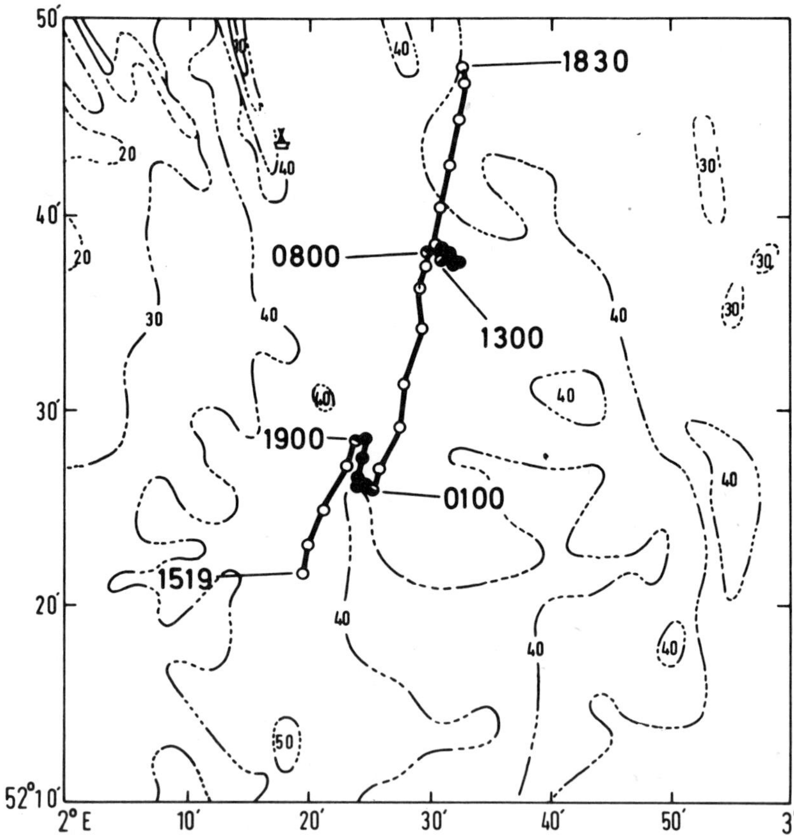

FIG. 14. Track chart of cod 4, 55 cm long, released from a cage at 1519 hours, 27 April 1975. Hourly positions of the fish are indicated and the times of slack water are given. ○, North-going tide; ●, south-going tide; ◓, low-water slack; ◒, high-water slack.

longer than might be expected. If his identification is correct and these echoes are returned by cod, it might be difficult to reconcile their movements with a maintenance of neutral buoyancy. But there are good reasons for believing that small cod can secrete and resorb gas faster than big cod; if the identification of the echoes is correct, could the fish be stratified by size, the small fish reaching the ceiling, the big fish being close to the floor?

Our measurements of gas secretion and resorption suggest that extensive vertical movements, if maintained for more than a few minutes, are incompatible with the maintenance of neutral buoyancy throughout the 24 hour period. As the gas-filled swimbladder makes a significant contribution to the target strength of a fish, these aspects of swimbladder physiology and fish behaviour should be of interest to those who use acoustic methods to estimate their abundance.

Dr. Greer Walker's tracking experiments with cod throw some light on these problems and the results of one of his tracking experiments show how laboratory and field observations can supplement each other.

A chart showing the movement of the cod over the ground and a plot of its depth during the track are given in Figs. 14 and 15, respectively. This fish, 55 cm in length, went north for 50 km in 27 hours at an average velocity

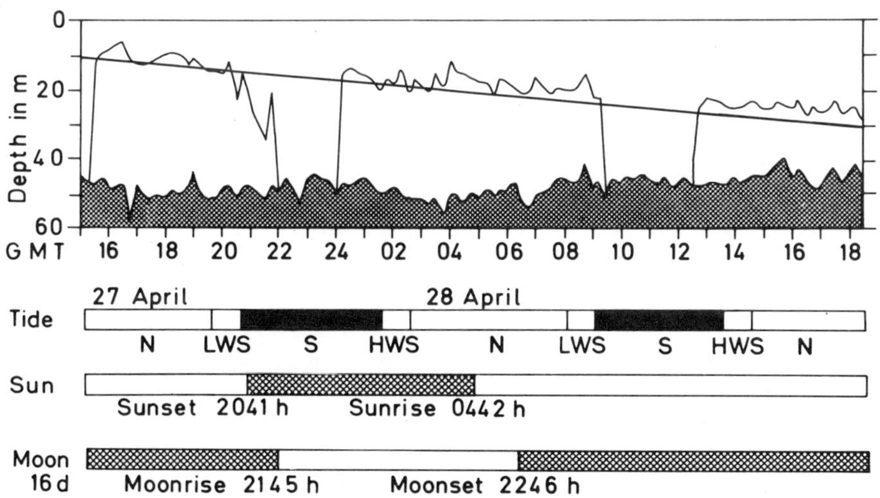

FIG. 15. The depth of Cod 4 in relation to the direction of the tide and other environmental factors. The cod was placed on the bottom in a cage at 1813 hours on 26 April 1975 and released at 1519 hours the next day. The continuous straight line indicates the depth at which the cod would be expected to be in neutral buoyancy assuming an initial equilibrium depth of 10 m and a gas secretion rate of 0·062 ml kg^{-1} minute^{-1} (equivalent to a descent rate of 0·7 m hour^{-1}).

of 50 cm s^{-1}, equivalent to $0{\cdot}9L\ s^{-1}$; it covered most ground when the tide was flowing in the overall direction of movement: like the plaice the cod appears to use selective tidal transport. These results will be published in detail (Greer Walker and Arnold, in preparation).

Let us see how the vertical movements of this cod fit in with the results of our laboratory experiments on gas secretion. The fish was left in a cage on the bottom at a depth of 45 m for 22 hours before being released with the use of a simple tripping device. (The estimate that we gave Dr. Greer Walker of the time that the cod should have been left in the cage was clearly wrong; it should have been kept in the cage for about 64 hours. But the consequences of the mistake were instructive.) The water temperature was 8 °C and the corresponding rate of gas secretion was equivalent to an adaptation rate of 0·7 m $hour^{-1}$. After 22 hours the fish should have been adapted to a depth of 15 m. When released from the cage the fish swam up to a depth of about 10 m, somewhat less than the depth at which it should have been neutrally buoyant. That was interesting. When the cod came up into midwater a second time, at 2400 hours, it reached a depth of about 15 m, and on the third occasion, at 1230 hours the following day, it reached a depth of 22 m. By and large there was a steady increase in the depth that the fish maintained when in midwater. The increase in depth was similar to that to be expected if the swimbladder was being filled with gas at a rate equivalent to the depth change of 0·7 m $hour^{-1}$, suggested by the laboratory experiments for 8 °C. Our laboratory experiments on gas secretion and resorption were undertaken, in part at least, to help with the tracking work at sea. I suspect that the tracking work will help us to interpret the laboratory studies.

We do not know how a fully adapted cod will behave in the sea. But the disparity between the rates of gas resorption and secretion should restrict semi-diurnal vertical migration to distances of only a few metres above the sea bed. If cod use a selective transport system, they would need to do no more than this to catch the tide and Dr. Greer Walker's track certainly suggests a decrease in vertical amplitude with time, that is, as adaptation proceeds. If this proves to be true it will provide a remarkable contrast with the plaice, it will be consistent with our ideas on buoyancy, and will support the hypothesis that the wise and energy-conscious fish should move quickly over the ground while swimming slowly through the water.

We have heard from Professor Lighthill and Dr. Wardle that fish swim in such a way as to reduce the power required. I conclude with the suggestion that some fish may have gone one step further and that they behave in such a way as to reduce the requirement to swim.

References

ALEXANDER, R. M. (1967). "Functional design in fishes." Hutchinson University Library, London, 160pp.

ANON (1965). Report of the working group on sole. *Co-op. Res. Rep.* **5**, 126pp.

BEAMISH, F. W. H. (1966). Vertical migration by demersal fish in the Northwest Atlantic. *J. Fish. Res. Bd Canad.* **23**, 109–139.

BOLSTER, G. C. (1955). English tagging experiments. *Rapp. P.-v. Réun. Cons. Int. Explor. Mer* **140** (2), 11–14.

BOLSTER, G. C. (1974). The mackerel in British waters. *In* "Sea Fisheries Research" (F. R. Harden Jones, ed.), pp. 101–116. Elek Science, London, 510pp.

BROCK, V. E. and RIFFENBURGH, R. H. (1960). Fish schooling: a possible factor in reducing predation. *J. Cons. Int. Explor. Mer* **25**, 307–317.

BRUNEL, P. (1972). The Gaspé cod ecosystem in the Gulf of St. Lawrence. III. The daily and seasonal vertical migrations of cod (*Gadus morhua*) in 1960–62. *Naturaliste Can.* **99**, 287–357.

CLEMENS, H. B. (1961). The migration, age, and growth of Pacific albacore (*Thunnus germo*), 1951–1958. *Fish. Bull. Calif.* (115) 128pp.

CUSHING, D. H. and HARDEN JONES, F. R. (1967). Sea trials with modulation sector scanning sonar. *J. Cons. Int. Explor. Mer* **30**, 324–345.

DICKSON, R. R. (1976). Field tests of seabed drifters using sector scanning sonar techniques. *J. Cons. Int. Explor. Mer* **37**, 3–15.

ELLIS, G. H. (1956). Observations on the shoaling behaviour of cod (*Gadus callarias*) in deep water relative to daylight. *J. Mar. Biol. Ass. U.K.* **35**, 415–417.

GREER WALKER, M., HARDEN JONES, F. R. and ARNOLD, G. P. (1977). The movements of plaice (*Pleuronectes platessa*) in the open sea. *J. Cons. Int. Explor. Mer* (in press).

GREER WALKER, M., MITSON, R. B. and STORETON-WEST, T. J. (1971). Trials with a transponding acoustic fish tag tracked with an electronic sector scanning sonar. *Nature, Lond.* **229**, 196–198.

HARDEN JONES, F. R. (1968). "Fish Migration." Arnold, London, 325pp.

HARDEN JONES, F. R., GREER WALKER, M. and ARNOLD, G. P. (1973). The movements of a Woodhead seabed drifter tracked by sector scanning sonar. *J. Cons. Int. Explor. Mer* **35**, 87–92.

HARDEN JONES, F. R., GREER WALKER, M. and ARNOLD, G. P. (in preparation). The migration of plaice (*Pleuronectes platessa*) in the Southern Bight of the North Sea.

HARDEN JONES, F. R. and MARSHALL, N. B. (1953). The structure and functions of the teleostean swimbladder. *Biol. Rev.* **28**, 16–83.

HARDEN JONES, F. R. and SCHOLES, P. (in preparation). The rates of gas secretion and resorption in the cod.

LOWNDES, A. G. (1937). The density of some living aquatic organisms. *Proc. Linn. Soc. Lond.* (150th session), Part 2, 62–73.

MATHER, F. J., MASON, J. M., SCHUCK, H. A. and JONES, A. C. (1977). Life history and fisheries of Atlantic blue fin tuna *Adv. Mar. Biol.* **14** (in press)

MITSON, R. B. and COOK, J. C. (1971). Shipboard installation and trials of an electronic sector-scanning sonar. *Radio Electron. Engr.* **41**, 339–350.

MITSON, R. B. and STORETON-WEST, T. J. (1971). A transponding acoustic fish tag. *Radio Electron. Engr.* **41**, 483–489.

NORMAN, J. R. (1963). "A History of Fishes", 2nd ed. Ernest Benn, London, 398pp.
VOGLIS, G. M. and COOK, J. C. (1966). Underwater applications of an advanced acoustic scanning equipment. *Ultrasonics*, **4**, 1–9.
WEIHS, D. (1973). Optimal fish cruising speed. *Nature, Lond.* **245**, 48–50.
WEIHS, D. (1977). Tidal transport as an efficient means of migration. *J. Cons. Int. Explor. Mer.* **37** (in press).

11. The Application of Large Amplitude Elongated Body Theory to Measure Swimming Power in Fish

C. S. WARDLE AND A. REID

Marine Laboratory, Aberdeen

1. Introduction

Details of the history of the methods developed for measurement of the power of swimming fish are reviewed by Webb (1975). The reactive model proposed as the large amplitude elongated body theory by Lighthill (1971) is applied in the present experiments to estimate the power output of large cod, *Gadus morhua*. A major advantage of the reactive theory is the economical way in which relatively few kinematic parameters are recorded from the image of a swimming fish in order to maintain a substantial degree of accuracy in representing the motions of the fish.

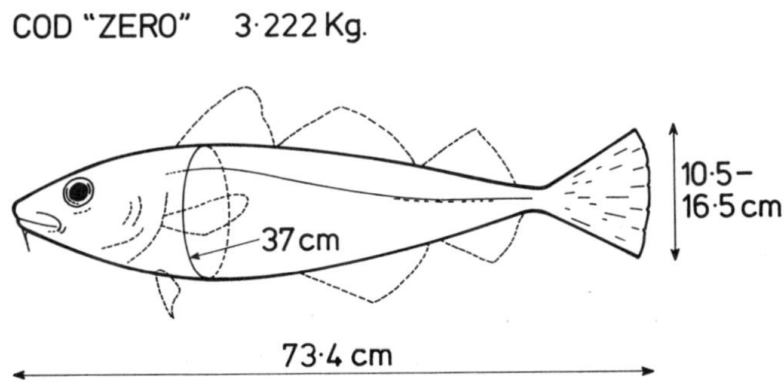

FIG. 1. Cod "Zero" was used as the test fish in the swimming power measurements. During faster swimming the fins shown dotted are collapsed against the body wall.

The cod was chosen as a fish having a subcarangiform mode of swimming (see Webb, 1975, Fig. 50). At slow swimming speeds of less than half a length per second the median fins are extended and the cod tends to swim as an anguilliform fish. As the swimming speed increases to beyond one length per second these median fins are folded (Fig. 1) leaving a narrow tail peduncle and a large triangular caudal fin and the subcarangiform swimming motion can then suitably be analysed by the reactive approach.

2. Methods

Methods for recording fish swimming movements

The circular gantry tank in the fish behaviour unit at the Marine Laboratory, Aberdeen, has a diameter of 10 m (Fig. 2). The filtered sea

FIG. 2. A general view of the experimental 8 m swimming track as arranged in the 10 m diameter gantry tank. The perspex raft is positioned at the centre of the race track with one TV camera in the ceiling (top centre) and the second TV camera viewing the fish from one side is hidden under black polythene on the right-hand edge of the tank. The fish swim between the far edge of the tank to the near edge where they are fed alternately by the assistants seen in the picture at the flashing lights (see text).

water is maintained at 12 °C and 90% oxygen saturation and is 1 m deep. Cod, *Gadus morhua*, caught off Girdleness, Aberdeen, in water 7–10 m deep, were selected and cold branded (Everest and Edmundson, 1967) with easily recognized patterns so that each individual could be identified during swimming experiments. The newly caught cod started to feed after 5 days in the tank and in the present series of experiments the group of 6 cod including the cod "Zero" used in the power analysis were fed at two feeding points 8 m apart. At each feeding point a filament lamp (12 V, 2·5 watts (W)) mounted in a transparent dome was arranged to flash on and off at 0·5 s intervals when required. At feeding time one of the flashing lights was switched on and food (chopped squid) was added within 20 cm of the light. As soon as all the food had been taken this light was switched off and the other light 8 m away was switched on at the moment food was added at that point. After 1 or 2 weeks a flashing light switched on at either position in the tank would attract all the fish directly to that point even if food was not always presented.

A reference grid of 25 cm squares was marked on a plastic sheet (2 × 1 m) coated with reflex reflector material (Scotchlite 3270) and placed on the floor of the tank half-way between the feeding lights. A television surveillance camera (Grundig FA42S fitted with a silicon diode array, vidicon tube and carefully set for correct geometry) was mounted vertically above the centre of the grid in the ceiling of the tank room. This camera formed a television picture of top views of fish against the grid seen through a floating, clear perspex raft which effectively eliminated surface water movements (Figs. 2 and 3). A second television camera was positioned at one side of the track and formed a side-view of the same fish showing information of swimming depth, fin usage and tail fin height against a second reference screen mounted vertically (Fig. 2). The two cameras were synchronized and the pictures from each were recorded on separate videotape recorders with the addition of a clock to identify frames and an electronic dot pattern as a grid to check geometry. Each camera was fitted with a 15 W, neon, strobed, light source as a ring around the lens. The light (duration 10 μs) was made to flash between the frame scans of the cameras and thus created synchronized frozen image pairs, at 0·020 s intervals, of top and side-views of a fish silhouetted against the reflex reflector reference grids.

With the television system operating the cod were made to race between the feeding points by flashing the appropriate feeding light. The voice of an observer simultaneously recorded on the videotape identified each fish as it crossed the reference grid. To analyse a selected series of frames, they were replayed singly on to a carefully adjusted television monitor scanning only the central flat portion of the screen and were copied on to 35 mm film. The 35 mm negative was then projected on to the sensitive table of the *X–Y* coordinate recorder (D-mac) and recorded on punched tape as indicated in the Appendix.

The analytical system described, with 50 frames per second, is only suited to large fish swimming at slower speeds. The shortest time in which a cod

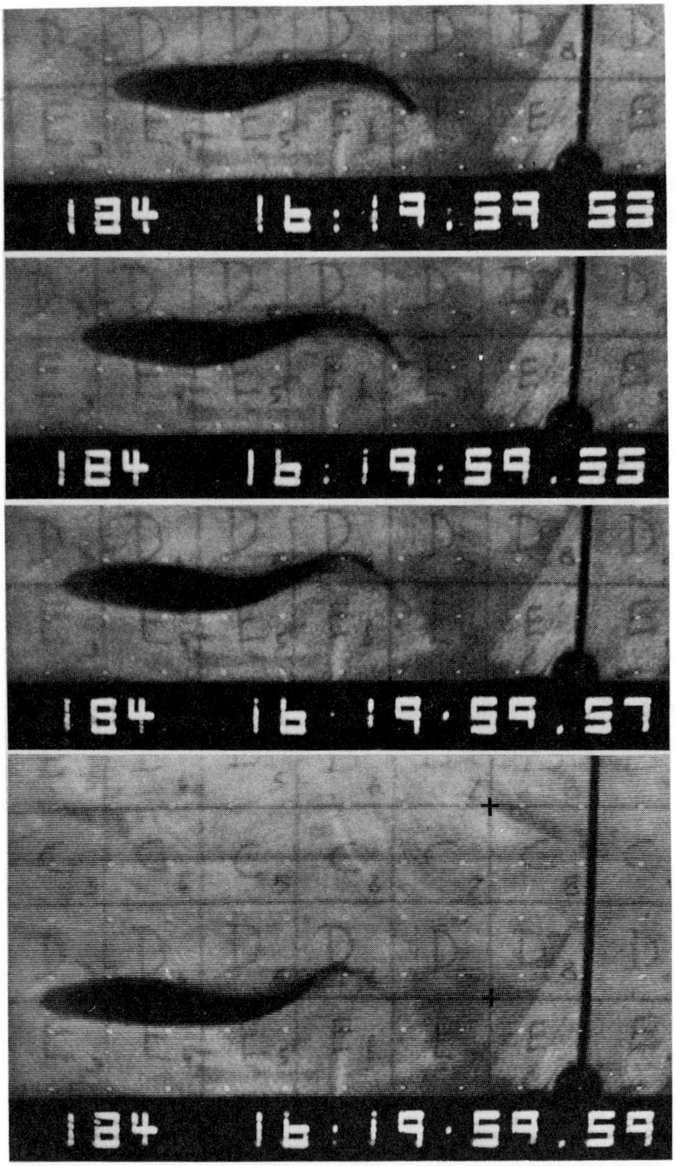

FIG. 3. Cod "Zero": part of one tail sweep during traverse 38. The right-hand clock figures read 9·53–9·59 s. The frozen image of the cod is seen as a silhouette against the background grid of squares (25 cm) through a perspex floating raft. The white dots are an electronic grid used to check the geometry of the TV system. In Fig. 8 (Nos. c12–15) are shown interpretations of head and tail positions taken from these pictures. The reference points marked + + are those used to scale the analysis.

73·4 cm long can completely contract one lateral muscle during swimming is near 40 ms (Wardle, 1975). This allows time for only two television frames, so that if this cod were to swim at maximum speed past the camera each tail beat would be recorded by only four frames. Using smaller fish at maximum swimming speeds (as shown in Wardle, 1975, Figs. 3 and 4), would result in only two frames per tail beat and 300 frames per second would be required to record sufficient information. The size of the tank limits the maximum speed of the large cod to 3 m s^{-1}, the maximum speed possible is predicted as near 5·5 m s^{-1} (Wardle, 1975).

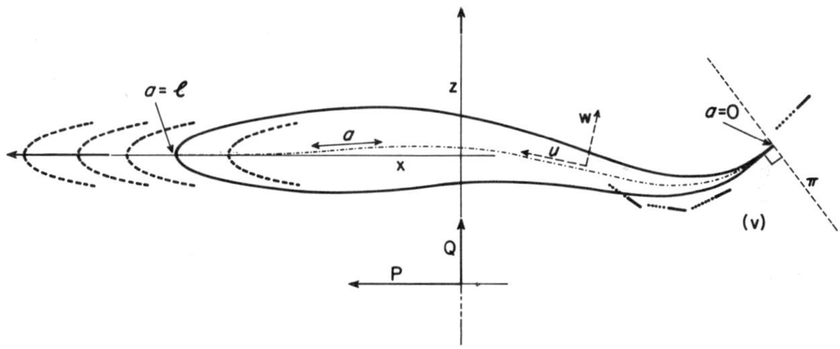

FIG. 4

Of the six cod racing between the lights in these experiments, the records of four crossings of one of them were chosen for careful study in this paper. Cod "Zero" (Fig. 1) weighed 3·222 kg had an overall length of 0·734 m, and a caudal fin height varying between 10·5 cm when relaxed to 16·5 cm when stretched. A natural stretched height of 0·015 m was used for the caudal fin height s in the calculations of virtual mass.

The mathematical analysis

The theory and worked example outlined by Lighthill (1971) define the measurements required. For the sake of completeness the essentials of the theory are outlined here where Lighthill's (1971) symbols are used throughout.

The position of any point on a fish swimming in a fixed horizontal plane is described by a coordinate system (x, z); let $l =$ length of the fish and define a "Lagrangian" coordinate a taking values from 0 to l along the backbone of the fish (Fig. 4). a thus defines a point on the spinal column by its distance from the tip of the fish's tail. To avoid the difficulties of considering the momentum changes in the fish's wake a vertical plane π, normal to the tip of the tail is drawn through the tail tip bounding the half space V containing the fish and excluding the wake. The assumptions are made that: (1) the spinal column does not change in length; (2) the fish has a uniform cross-section and hence lateral symmetry of water momentum. Lighthill (1971) derives the vector equation for the rate of change of momentum within V

$$\frac{\mathrm{d}}{\mathrm{d}t}\int_0^l mw\left(-\frac{\partial z}{\partial a},\frac{\partial x}{\partial a}\right)\mathrm{d}a = \left[-umw\left(-\frac{\partial z}{\partial a},\frac{\partial x}{\partial a}\right)+\tfrac{1}{2}mw^2\left(\frac{\partial x}{\partial x},\frac{\partial z}{\partial a}\right)\right]_{a=0} - (P,Q), \quad (1)$$

where at any point along the spinal column u denotes the motion of the fish tangential to the spine, w the motion of the fish normal to the spine (Fig. 4), m the virtual mass (see Fig. 5) and the vector (P, Q) denotes the reactive force with which the water acts on the fish.

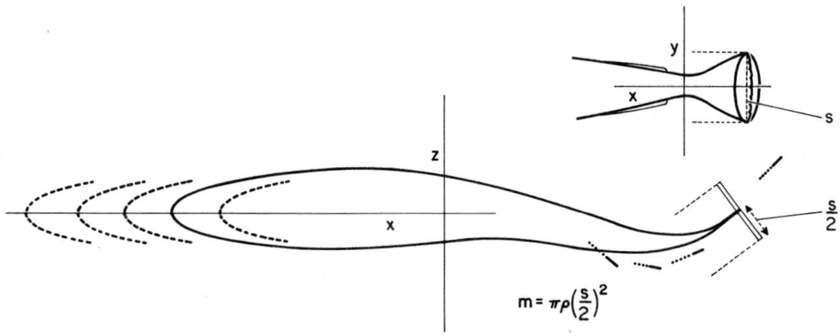

FIG. 5

Equation (1) is concerned purely with momentum calculated from a reactive point of view and neglects any influence of the viscous boundary layer at the fish surface. The left-hand side (equation (1)) denotes the total rate of change of water momentum within V; on the right-hand side in the square brackets the first term defines the rate of change of momentum due to the motion of π while the second added term denotes the rate of change of momentum due to pressure forces acting across π. From another point of view equation (1) states:

The force exerted by the fish on the water = Total rate of change of momentum within V − Rate of change of momentum V caused by its boundary π and not by the fish.

Equation (1) simplifies to

$$(P,Q) = \left[mw\left(\frac{\partial z}{\partial t},-\frac{\partial x}{\partial t}\right)-\tfrac{1}{2}mw^2\left(\frac{\partial x}{\partial a},\frac{\partial z}{\partial a}\right)\right]_{a=0} - \frac{\mathrm{d}}{\mathrm{d}t}\int_0^l mw\left(-\frac{\partial z}{\partial a},\frac{\partial x}{\partial a}\right)\mathrm{d}a \quad (2)$$

(Lighthill, 1971, equation (6)) being a formula for the instantaneous thrust exerted by the fish both in the x and z directions. The longitudinal thrust P

and the side-force Q are then

$$P = \left[mw\frac{\partial z}{\partial t} - \tfrac{1}{2}mw^2\frac{\partial x}{\partial a}\right]_{a=0} + \frac{\mathrm{d}}{\mathrm{d}t}\int_0^l mw\frac{\partial z}{\partial a}\,\mathrm{d}a, \tag{3}$$

$$Q = \left[-mw\frac{\partial x}{\partial t} - \tfrac{1}{2}mw^2\frac{\partial z}{\partial a}\right]_{a=0} - \frac{\mathrm{d}}{\mathrm{d}t}\int_0^l mw\frac{\partial x}{\partial a}\,\mathrm{d}a. \tag{4}$$

To return to equation (3) it is clear that to evaluate the final integral term and subsequently its rate of change would require detailed knowledge of the curvature of the spinal column. However, provided the fish swims in the x direction and moves in a fashion symmetrical about the x-axis then this integral

$$\int_0^l mw\frac{\partial z}{\partial a}\,\mathrm{d}a$$

will be periodic and so its rate of change over a symmetrical tail cycle will be zero. It is therefore practical to calculate the mean longitudinal thrust $\bar{P}$ produced by the fish as

$$\bar{P} = \overline{\left[mw\frac{\partial z}{\partial t} - \tfrac{1}{2}mw^2\frac{\partial x}{\partial a}\right]_{a=0}}. \tag{5}$$

Having selected a suitable set of pictures of a swimming fish then for each picture, P_i is evaluated in

$$P_i = \left[mw\frac{\partial z}{\partial t} - \tfrac{1}{2}mw^2\frac{\partial x}{\partial a}\right]_{a=0} \tag{6}$$

and then for the series of pictures the mean thrust

$$\bar{P} = \frac{1}{n}(P_{i_1} + P_{i_2} + P_{i_3} + \ldots + P_{i_n}). \tag{7}$$

Values for the virtual mass m (equation (5)) are calculated as discussed in Lighthill (1970) where

$$m = \pi\rho\left(\frac{s}{2}\right)^2 \quad \text{(Fig. 5)}$$

(ρ is the density of the water and s is the tail height). Values for the partial derivatives $\partial x/\partial a$ and $\partial z/\partial a$ are obtained by measuring the angle θ that the tangent to the tail at the tail tip makes with the x-axis and then

$$\cos\theta = \frac{\partial x}{\partial a} \quad \text{and} \quad \sin\theta = \frac{\partial z}{\partial a} \quad \text{(Fig. 6)}.$$

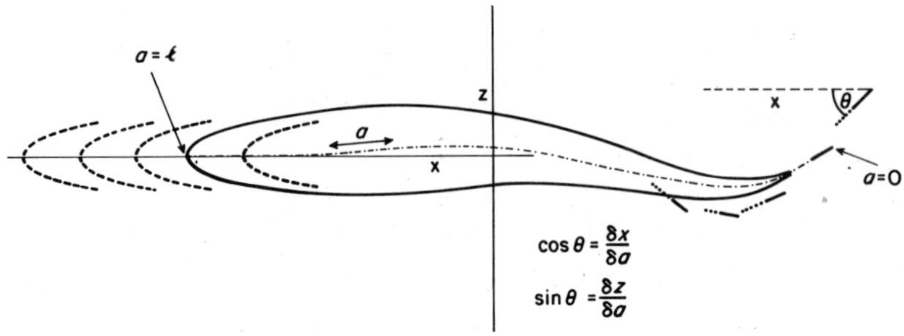

FIG. 6

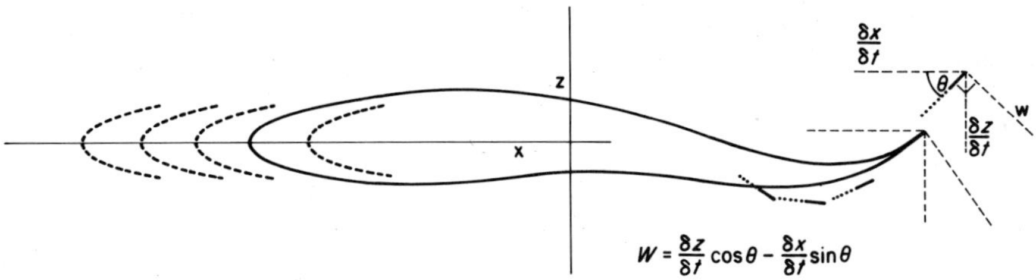

FIG. 7

The tail tip velocity

$$w = \frac{\partial z}{\partial t} \cos \theta - \frac{\partial x}{\partial t} \sin \theta \quad \text{(Fig. 7).}$$

The Application of the Analysis

A summary of the computer program FISHGO is given in the Appendix and the process of recording the coordinates using the D-mac followed the routine pattern outlined. Constant dimensions of fish length, grid points and tail height are typed in and then from each image of the fish was recorded the position of the two reference points from the background grid, a head tip position, a tail tip position and a second tail blade position to give the angle of the caudal fin θ.

The position coordinates of the fish were first scaled by using the two reference points from the background grid, but as the fish was swimming, midwater, above the reference grid, a correction to this scale was made by measuring the average length between the head and tail tip coordinates. The factor required to correct the apparent length of the fish to its known true length (see Table 1) was then determined and used to correct all spacings of the other coordinates in the swimming plane of the fish.

In the initial version of the computer program the direction in which the head moved from frame i to $i+1$ was used as the x-axis in the analysis of

TABLE 1. Determination of the scale correction factor, the table shows the head tip and tail tip coordinates and the estimated length in each frame of cod "Zero" swimming mid-water above the reference grid

FRAME NO.	HEAD X	HEAD Z	TAIL X	TAIL Z	FISH LENGTH
1	-0.270	.0047	0.479	.0163	.8248
2	-0.319	.0023	0.419	.0438	.8141
3	-0.377	.0079	0.362	.0666	.8163
4	-0.433	.0019	0.303	.0601	.8126
5	-0.488	-.0034	0.240	.0278	.8024
6	-0.542	-.0055	0.187	-.0165	.8019
7	-0.595	-.0013	0.132	-.0507	.8013
8	-0.647	-.0102	0.062	-.0559	.7805
9	-0.701	-.0015	0.013	-.0089	.7851
10	-0.760	.0045	-0.038	.0460	.7957
11	-0.820	-.0042	-0.107	.0711	.7904
12	-0.861	-.0016	-0.168	.0733	.7687
13	-0.911	.0029	-0.206	.0064	.7760
14	-0.972	-.0018	-0.260	-.0470	.7840
15	-1.024	-.0009	-0.327	-.0678	.7686
16	-1.078	.0035	-0.380	-.0532	.7681
17	-1.161	-.0027	-0.427	.0079	.8085
18	-1.227	-.0054	-0.487	.0557	.8197
19	-1.292	.0025	-0.539	.0844	.8366
20	-1.356	.0067	-0.591	.0788	.8486
21	-1.428	.0017	-0.644	.0434	.8653

AVERAGE FISH LENGTH = .8033 ACTUAL FISH LENGTH = 0.734

frame i. However, the significant yaw (up to ± 10 cm in the z-direction) of the tip of the head gave a continuously changing x-axis and produced misleading values for the tail coordinate z and inaccurate values of the tail tip speed $\partial z/\partial t$. To obtain a more consistent x-axis, all the head tip coordinates were examined at the start of the program and the best straight line through these points was calculated and taken to define the x-axis.

Initially the motion of the tail tip in frame i was defined by assuming it moved in a straight line from its position in frame i to its position in frame $i+1$, but it was soon realized that realistic values were not being obtained and subsequently numerical differentiation was used to evaluate the partial derivatives $\partial x/\partial t$, $\partial z/\partial t$. This made use of tail tip positions from the two preceeding frames and the two subsequent frames to calculate the motion of the tail tip at the instant frame i was taken. With the limited frame rate of only 50 per second, analysis of the rapid movements of the tail tip in faster swimming is only just practicable. As indicated earlier if the higher tail beat frequencies of smaller fish are to be analysed at their faster swimming speeds frame rates as high as 300 per second will be required, even for this minimum information.

It was intended (in the present experiments) to measure the value of the tail height s from each synchronized frame of the side-view of the swimming fish. However, as the changes in s were small in relation to the apparent changes in height due to the large movement towards and away from the side-view camera it was found impractical. In the cases analysed the fixed value of 0·015 m was used in the calculation of virtual mass m, and the program was arranged for a single s value.

In the frozen television images the interpretation of the position of the tangent to the tail tip in each frame was the most serious source of error particularly in those frames where the tail changes its direction at the end of a sweep. In practice the tail is not a flat plate but in the cod is effectively a flexible membrane edged by two rigid rods. The membrane takes up a natural curve as the tail sweeps with the two rigid rods leading and has the effect of producing a thick tail image (Fig. 3). It is therefore necessary to make a careful interpretation of the position of the tail tip and angle of the tail blade in each frame. As a guide to the effect on power output of an error in this judgement, $\partial P_i/\partial \theta$ was calculated for each value of P_i giving a value for the change in P_i for a 1 degree change in θ. Values could be as high as 3 newtons per degree (Table 2, DPDTHETA, frame 13). At turning points different assessments of θ could vary possibly as much as 40°. Thus we had to design a consistent method of interpretation and carefully follow the same pattern through a sequence of frames. Between the turns, the main driving sweep of the tail tip (Fig. 3) gave a clear series of angles and tail tip positions and allowed a fair degree of confidence in the values for P_i.

From careful examination of the cod tail it is clear that because of the flexibility of the driving surface there is a good deal of free play in the tail membrane at the turn where the driving force becomes zero on the tail membranes. For a brief period the membrane becomes slack before being pressed across in the reversed direction and this detail is not allowed for in the measurement at the turning points where instantaneous negative peaks are seen in the results (P, Figs. 9 and 10).

Distortion in the optical and television system was a variable feature (of the system used) but was carefully monitored by including both a physical reference grid (black lines Fig. 3) and an electronic grid (white dots Fig. 3), injected to the television picture. In attempts to rectify the effect of any distortion a program was written into which were input the coordinates of a selection of the grid points and the program was made to calculate a simple transformation which would restore any distorted grid points to a regular rectangular grid. The coefficients of these transformations were then input to the fish swimming program FISHGO and used to transform all

TABLE 2. Computer output of power analysis for traverse 38·2 for cod "Zero" (see text)

FRAME NO.	HEAD SPEED (M/S)	TAIL DXDT (M/S)	TAIL DZDT (M/S)	TAIL ANGLE THETA (DEG)	TAIL W (M/S)	TAIL DXDA	MWDZDT (N)	M*W*W* DXDA/2 (N)	THRUST P (N)	DPDTHETA (N/DEG)
1	1.900	2.994	1.374	24.505	0.008	0.910	0.1985	0.0005	0.1980	-1.424
2	2.953	1.432	0.570	20.013	0.046	0.940	0.4738	0.0179	0.4560	-0.257
3	2.897	2.899	0.495	-2.868	0.639	0.999	5.7367	3.7008	2.0359	0.127
4	2.764	3.087	-1.040	-33.817	0.854	0.831	-16.1021	5.4920	-21.5942	1.676
5	2.700	2.881	-2.065	-36.454	0.051	0.804	-1.9154	0.0191	-1.9345	2.361
6	2.675	2.599	-2.134	-24.409	-0.869	0.911	33.6297	6.2367	27.3930	1.331
7	2.629	3.246	-1.160	-6.293	-0.797	0.994	16.7704	5.7281	11.0423	0.379
8	2.611	3.032	1.132	33.801	-0.746	0.831	-15.3089	4.1891	-19.4981	-1.697
9	2.824	2.337	2.890	32.195	1.200	0.846	62.8890	11.0510	51.8380	-1.965
10	3.084	3.044	2.129	18.289	1.066	0.949	41.1420	9.7808	31.3612	-1.201
11	2.509	3.418	0.848	-0.875	0.900	1.000	13.8354	7.3417	6.4936	0.054
12	2.158	2.378	-1.771	-50.145	0.690	0.641	-22.1645	2.7682	-24.9327	1.962
13	2.833	2.125	-3.432	-41.203	-1.182	0.752	73.5690	9.5342	64.0348	2.959
14	2.864	3.150	-1.945	-18.785	-0.827	0.947	29.1781	5.8740	23.3040	1.292
15	2.490	3.105	-0.213	6.802	-0.579	0.993	2.2352	3.0191	-0.7839	-0.344
16	3.508	2.384	2.095	42.773	-0.082	0.734	-3.1047	0.0445	-3.1492	-2.163
17	3.860	2.659	2.994	31.643	1.154	0.851	62.6432	10.2763	52.3669	-2.331
18	3.207	2.867	2.000	25.699	0.559	0.901	20.2587	2.5504	17.7083	-1.612
19	3.191	2.568	0.622	-6.728	0.918	0.993	10.3507	7.5917	2.7590	0.212
20	2.968	2.624	-1.025	-35.198	0.675	0.817	-12.5477	3.3776	-15.9253	1.323
21	4.024	2.659	-1.769	-33.751	0.006	0.831	-0.1944	0.0003	-0.1947	1.793

VIRTUAL MASS (KG/M) 18.131 TAIL HT.S (M) .150

the fish head and tail coordinates. It was found that these transformations made little difference to the values of x and P as calculated by FISHGO.

3. Results

An example of the recorded coordinates of cod "Zero" during a single traverse is shown in Figs. 3 and 8. A full analysis of this example is shown as computer output in Table 2. In Fig. 9 some of the values are plotted along the horizontal time scale and the quantities θ, w, $\partial x/\partial t$ and P are all seen to be periodic, having a wavelength much the same as that of z, the tail tip coordinate. The instantaneous peaks of P_i correspond to the movements

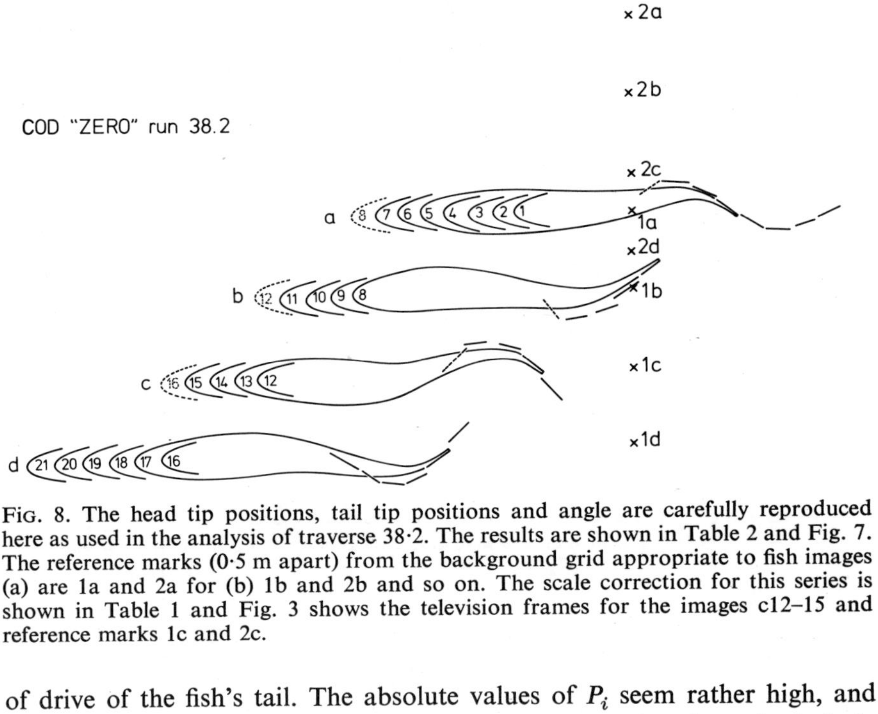

FIG. 8. The head tip positions, tail tip positions and angle are carefully reproduced here as used in the analysis of traverse 38·2. The results are shown in Table 2 and Fig. 7. The reference marks (0·5 m apart) from the background grid appropriate to fish images (a) are 1a and 2a for (b) 1b and 2b and so on. The scale correction for this series is shown in Table 1 and Fig. 3 shows the television frames for the images c12–15 and reference marks 1c and 2c.

of drive of the fish's tail. The absolute values of P_i seem rather high, and although in theory we had selected a series of pictures that showed relatively symmetrical tail movements, in practice this was never exactly so. Thus our values for P_i must contain to some degree an error caused by the mean of the term $(\mathrm{d}/\mathrm{d}t)\int_0^1 mw\,(\partial z/\partial a)\,\mathrm{d}a$ which had been assumed to be zero. Lighthill (1971) suggested a crude approximation for this integral of $[\frac{1}{2}mwz]_{a=0}$. In the example (Fig. 9) values were calculated and the rate of change of this quantity was found to be almost exactly complementary to P_i (Fig. 9, top). This result suggests that the peaks in P_i will be evened out and the fish

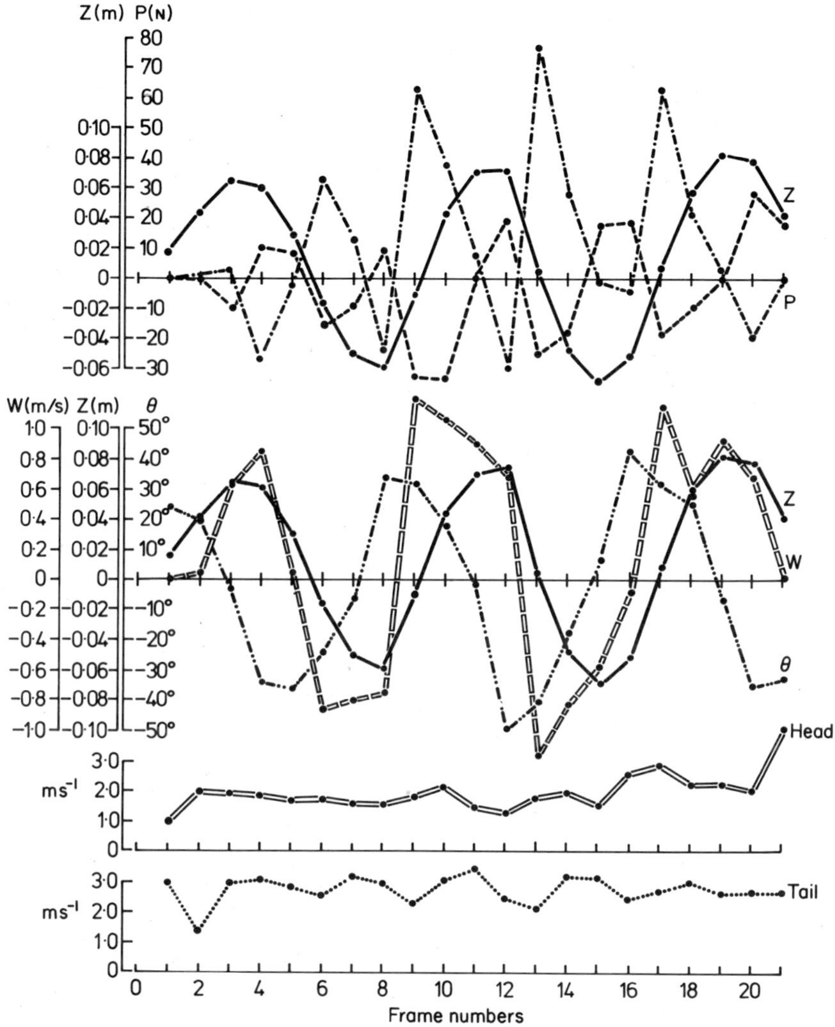

FIG. 9. Cod "Zero": traverse 38·2 shows the relationship between the different parameters of the tail beat as calculated by FISHGO computer program. The horizontal scales are all indicating time 0·02 s per division or frame number. In this traverse velocity increased from 2·7 to 3·0 m s^{-1}. The unlabelled dashed line in the top figure is $[\frac{1}{2}mwz]_{a=0}$.

produces a relatively more even value of the thrust throughout its swimming motion.

Three other traverses by the same fish at slightly different speeds are shown in Fig. 10, where for comparison P and z are plotted along the time axis. Duplicate interpretations of the photographs were made for two of the

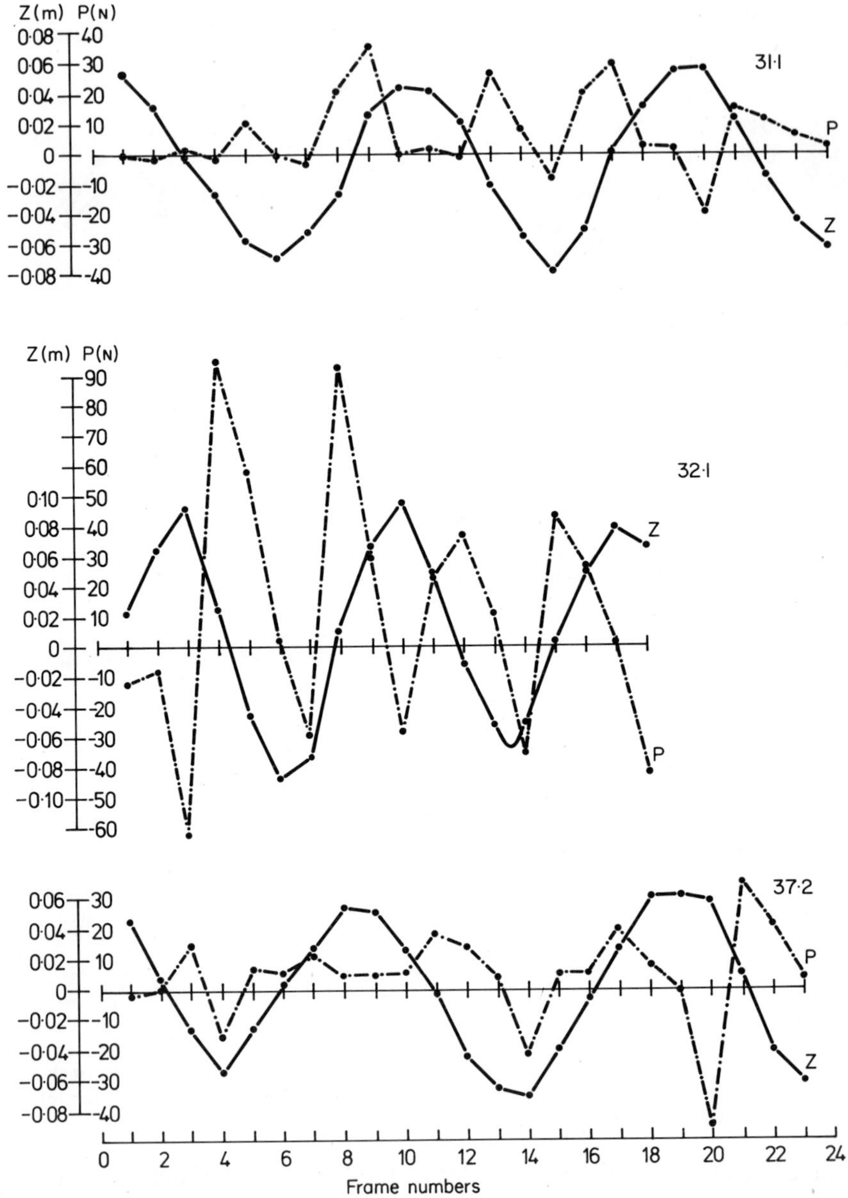

FIG. 10. Cod "Zero": three traverses showing tail tip position (continuous line) and instantaneous power (dotted line). The horizontal scale is time or frame number (0·020 s per division). (a) Traverse 31·1: velocity increases 1·9 to 2·1 m s^{-1}; (b) traverse 32·1: velocity increases 2·8 to 3·0 m s^{-1}; (c) traverse 37·2: velocity steady at 2·6 m s^{-1}.

traverses, the groups of figures used to derive the mean power output $\bar{P}\bar{V}$ are shown in Table 3 and power against velocity is plotted (Fig. 11) for the four traverses.

TABLE 3. Selection of frames used to calculate velocity $\bar{V}$, force $\bar{P}$ and power $\bar{P}\bar{V}$ plotted in Fig. 11. The weight of cod "Zero" was 3·222 kg. The data and frame numbers are the same as those used in Figs. 9 and 10

Traverse No.	No. of frames	Frame Nos.	Elapsed time (s)	$\bar{V}$	$\bar{P}$	$\bar{P}\bar{V}$	W kg^{-1}
31·1	10	1–10	0·200	1·88	6·43	12·08	3·7
	11	3–13	0·260	1·93	8·56	16·55	5·1
	10	6–15	0·300	1·91	8·29	15·83	4·9
	10	8–17	0·340	1·97	13·21	25·98	8·1
	10	10–19	0·380	2·01	7·98	16·08	5·0
	10	12–21	0·420	2·10	7·28	15·28	4·7
	10	15–24	0·480	2·07	5·88	12·18	3·8
32·1	8	3–10	0·200	2·85	19·63	55·91	17·4
	9	4–12	0·240	2·88	31·16	89·76	27·9
	8	6–13	0·260	2·93	17·32	50·73	15·7
	8	8–15	0·300	3·02	21·74	65·70	20·4
	8	10–17	0·340	2·97	9·78	29·06	9·0
37·1	10	2–11	0·220	2·56	5·34	13·66	4·2
	10	5–14	0·280	2·47	4·34	10·71	3·3
	11	6–16	0·320	2·48	5·11	12·65	3·9
	11	9–19	0·380	2·46	3·32	8·16	2·5
37·2	10	2–11	0·220	2·54	5·53	14·06	4·4
	11	6–16	0·320	2·55	5·39	13·76	4·3
	11	8–18	0·360	2·53	6·32	15·99	5·0
	11	11–21	0·420	2·54	4·04	10·28	3·2
38·1	9	1–9	0·180	2·70	8·13	21·98	6·8
	9	4–12	0·240	2·71	9·82	26·59	8·3
	9	6–14	0·280	2·71	16·89	45·76	14·2
	8	8–15	0·300	2·71	16·79	45·50	14·1
	9	9–17	0·340	2·91	21·96	63·88	19·8
	9	12–20	0·400	2·96	17·74	52·43	16·3
38·2	9	3–11	0·220	2·74	10·24	28·09	8·7
	9	5–13	0·260	2·67	16·76	44·72	13·9
	8	8–15	0·300	2·67	17·10	45·69	14·2
	9	11–19	0·380	2·96	15·87	46·93	14·6

The measured mean power output by cod "Zero" was between 10 W kg^{-1} and 30 W kg^{-1} at swimming speeds between 2·0 and 3·0 m s^{-1}. Three of the traverses (Table 3) show the cod accelerating during the 0·4 s. In traverse 31·1 cod "Zero" accelerated from 1·9 to 2·1 m s^{-1} and power output was between 20 and 65 W. In traverse 32·1 acceleration was from 2·8 to 3·0 m s^{-1} and used power between 30 and 90 W. In the duplicate analysis of traverse 37 the lowest power of between 8 and 16 W was measured at a steady speed of 2·6 m s^{-1}.

4. Discussion

The maximum power output of 90 W or 30 W kg^{-1} measured in these preliminary experiments with cod "Zero" occurred just before the maximum recorded swimming speed of 3·0 m s^{-1} (Table 3, run 32·1). As shown on Fig. 11, the other power measurements were lower and the lower values corresponded to lower speeds.

A summary of the values of swimming power obtained by various devices giving direct measurement is reviewed by Webb (1975, Table 6). All the values presented by Webb are low in relation to those measured in cod "Zero" except those for *Sphyraena barracuda* caught on a line where values of 34·5 W kg^{-1} fish weight and 66·5 W kg^{-1} are quoted (after Gero, 1952). Gero's *S. barracuda* No. 24, weight 9·1 kg, length 1·22 m with 1·8 kg muscle, showed a peak speed of 12 m s^{-1} and at this speed power output was estimated between 49 W kg^{-1} and 66 W kg^{-1} (fish weight) with 50% to 75% laminar flow over the body; however, an estimate of the power required was nearer 150 W kg^{-1} if flow was 100% turbulent (from Gero, 1952, Fig. 18).

Current knowledge of the biochemistry of swimming in fish allows a prediction to be made of which types of swimming muscle are being used at the different swimming speeds. In a paper discussing this subject Wardle (1977) reviews the evidence which suggests that teleosts swimming aerobically have their sustained swimming speed limited by their overall oxygen supply. The maximum rate of uptake of oxygen by fish in the family Gadidae was measured in a typical representative the haddock, *Melanogrammus aeglefinus* (length 0·248 m), by Tytler (1969) at 300 mg of oxygen kg^{-1} $hour^{-1}$ while cruising at 0·5 m s^{-1}. It is known that 1 mg of oxygen releases approximately 20 joules (J) of energy so that the cod might be able to release 6000 J kg^{-1} $hour^{-1}$ or 1·67 W (J s^{-1}) per kg of fish weight. Returning to cod "Zero" (weight 3·22 kg) this would set a limit to aerobic power output of 5·4 W.

It is likely that the rested anaerobic muscle of teleosts and mammals will have similar power output when measured at their temperature of

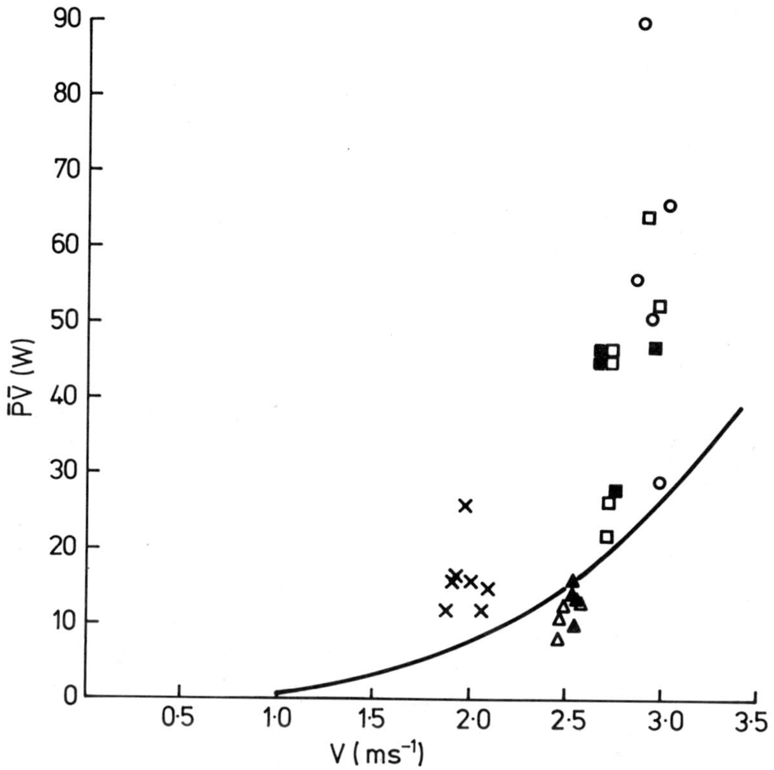

FIG. 11. Cod "Zero": showing mean power output against mean velocity. The values were determined by appropriate grouping as shown in Table 3. Key to traverse numbers: ×—31·1; ○—32·1; △—37·1; ▲—37·2; □—38·1; ■—38·2. The continuous line shows $y = x^3$.

adaptation. It is apparent from existing measurements of the levels of activity of the glycolytic enzymes in muscle at physiological temperatures that cold water poikilotherms compensate for the lower temperatures of their tissues by an adaptation which increases the catalytic effect of their glycolytic enzymes (Hochachka and Somero, 1971). The anaerobic fuel, glycogen, is present in cod muscle as in most teleost fish to levels near 1 g per 100 g of muscle, each gramme of glycogen when converted into lactic acid by anaerobic glycolysis releases 558 J. In the 1·52 kg (muscle weight of cod $= 3{\cdot}75 \times L^{2{\cdot}86}$ (Wardle, 1977)) of swimming muscle of cod "Zero" when completely rested there could be present a store of 8481 J of anaerobic energy. One important property of this store is that its rate of release into muscle power for movement appears to be limited only by the contraction time and strength of the muscle. It is therefore quite practical to consider the whole amount being released at 160 W in 53 s, at 80 W during 106 s, or 40 W during 212 s and so on. However, it should be noticed that as the glycogen is used the endurance and probably the strength of the muscle used in high speed swimming are reduced accordingly. Teleosts have behaviour patterns that lead to conservation of the anaerobic store as without it they become exceedingly vulnerable. The diminishing power output of a fish caught on a fishing line is familiar to the angler.

Earlier estimates of the power used during swimming have been made on entirely separate principles. Bainbridge (1961) derived a formula equating power necessary for fish locomotion to power available for fish locomotion. His approach is particularly relevant as it allowed an estimate of the power required for maximum speed swimming. Bainbridge developed the use of the standard hydrodynamic formula (after Gray, 1936) giving the drag force of a surface where Drag $= \frac{1}{2}\rho A V^2 C_f$, where ρ is the density of water, A the surface area of the fish, V the velocity of movement and C_f the coefficient of frictional drag. He corrected this for a three-dimensional body by including form or pressure drag allowing for the fishes fineness ratio after Hoerner (1965, p. 6–17, formula 28), where

$$C_{D\,(\mathrm{wet})}/C_f = 1+15(d/L)^{\frac{3}{2}}+7(d/L)^3.$$

$C_{D\,(\mathrm{wet})}$ is the total drag due to wetted surface area and for a variety of specimens substitution of d (diameter) and L (length) gives a form drag coefficient close to $1{\cdot}2C_f$:

$$\text{Drag force} = \tfrac{1}{2}\rho A V^2 \times 1{\cdot}2C_f.$$

The power necessary to propel such a fish is this drag force times the velocity

$$= \tfrac{1}{2}\rho A V^3 \times 1{\cdot}2C_f.$$

Bainbridge estimated the power available in half the muscle mass (W_m) of the fish assuming a 75% efficiency so that at any velocity

$$\tfrac{1}{2}\rho A V^3 \times 1{\cdot}2C_f = \tfrac{3}{4}P(W_m/2).$$

The area of the fish is found to be near $0{\cdot}4L^2$ so that

$$0{\cdot}64L^2\,V^3\,C_f = PW_m.$$

The drag (C_f) varies with laminar or turbulent flow which is then considered by including Reynolds number (Re) (i.e. length L times velocity V divided by kinematic viscosity of the medium). The laminar and turbulent drag coefficients are $1{\cdot}328\mathrm{Re}^{-0{\cdot}5}$ and $0{\cdot}074\mathrm{Re}^{-0{\cdot}2}$. A discussion of the state of turbulence at burst speed swimming is included in Wardle (1977) where it was argued that a fish of the cod type and size (swimming at speeds greater than sustained cruising speed) is having to cope with a considerable degree of turbulence at Reynolds numbers greater than 1×10^6 (Fig. 12). If we consider drag with surface flow in a condition of turbulence

$$\text{the turbulent velocity} = \left[\frac{PW_m}{L^{1\cdot 80}}\right]^{0\cdot 357} \times 0\cdot 677 \ (\mathrm{m\,s^{-1}}).$$

(Note Bainbridge (1961) uses cgs units and this is the same formula in SI units.) W_m, as we have seen for large cod above, was measured as $3\cdot 75 \times L^{2\cdot 86}$ (kg). Bainbridge for his smaller fish found W_m to be $3\cdot 155 \times L^{2\cdot 9}$ (kg).

$$V_{\mathrm{turb}} \text{ was thus written:} \quad P^{0\cdot 357} \times L^{0\cdot 393} \times 1\cdot 020.$$

By substituting suitable values for P, where L and maximum V were known, Bainbridge was able to show that most observed maximum swimming speeds could be explained by a maximum power output of near 40 W kg^{-1}

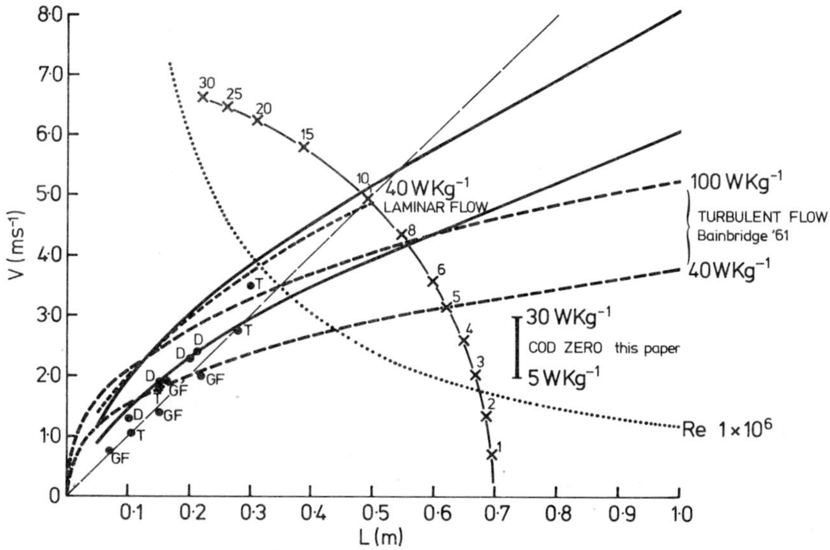

FIG. 12. The power estimates and measurements discussed in the text are here related to the swimming speed V and fish length L. The swimming power measured in cod "Zero" is shown as a short vertical line at $L = 0\cdot 73$ m. The dashed lines are derived from Bainbridge's (1961) calculations for power needed in laminar and turbulent flow. The letters indicate Bainbridge's maximum measured swimming speeds for: D, dace; GF, goldfish; T, trout. The firm lines indicate predicted maximum swimming speeds at 14 °C and stride lengths 0·6–0·8 derived from measurements of muscle contraction time (Wardle, 1975, 1977). The critical transition between laminar and turbulent flow (for smooth bodies) is shown by the iso-Reynolds number line Re 1×10^6. The crosses forming a third scale when connected to the origin indicate body lengths per second, a line at 10 bL s^{-1} is marked.

for a laminar flow over the swimming fish, which might include those fish less than 30 cm studied by Bainbridge or near 100 $W kg^{-1}$ for swimming at maximum speed, with turbulent flow over the skin surface, e.g. in the 73·4 cm cod "Zero" at 5·5 $m s^{-1}$. A summary of this finding is outlined in Fig. 12. Also shown are points representing maximum swimming speeds of some freshwater fish measured by Bainbridge (Fig. 12, capital letters); the maximum swimming speeds predicted by Wardle (1975, 1977) from contraction time of swimming muscle of marine fish; and for comparison the power and speed of cod "Zero" (Fig. 12, short vertical line at 0·73 m). On the same figure is shown for reference the iso-Reynolds number line for 1×10^6.

The direct application of the reactive theory to measure swimming power in fish appears to give results which are in agreement with both biochemical assessment of power available and estimates based on drag to be overcome during swimming. Recording power output by direct measurement has not been generally successful due to the interference with the performance of the fish but here again existing measurements have been shown to reach a similar order of magnitude. Even when physical interference is absent great care is required to avoid damaging, inhibiting or restricting the performance of the fish by transferring them to the laboratory. In the experience of the authors it is relatively rare for fish to swim in a uniform manner which can be analysed, but the use of videotape to record the swimming of fish has increased the chances of recording good examples of swimming when these do occur. Further measurement of fish power output using modern hydrodynamic theory would be worth-while in a variety of species at different swimming speeds. Measurements of the power available in fish swimming muscle and how this power is transmitted to the tail fin of the fish would supply a valuable check of the power limit available to a fish.

Appendix

OUTLINE OF PROGRAM FISHGO

INPUT

1. Distance between background reference points.
2. Tail height, *s*.
3. Fish identification numbers.
4. Number of frames, *N*, in the film sequence being analysed.

5. For each of the N frames, 5 pairs of (x, y) coordinates are input:
   coordinates of two background reference points
   head tip coordinates
   tail tip coordinates
   tail direction coordinates
6. Fish length.

The data is input and coordinates transformed to a system consistent from frame to frame. Corrections for distortion and magnification are applied. The best straight line through the fish head coordinates is calculated and then the coordinate system is transformed to one having this line as the X-axis. A tail angle for each frame is calculated.

Partial derivatives with respect to time are calculated numerically and the Lighthill's formula evaluated as indicated in the text.

Finally, a table of results is output as shown in Table 2.

A more detailed explanation of the program and a copy of the print-up may be obtained from the authors.

References

Bainbridge, R. (1961). Problems of fish locomotion. *Symp. Zool. Soc. Lond.* **5**, 13–32.

Everest, F. H. and Edmundson, E. H. (1967). Cold branding for field use in marking juvenile salmonids. *Progve. Fish. Cult.* **29**, 175–176.

Gero, D. R. (1952). The hydrodynamic aspects of fish propulsion. *Am. Mus. Norit.* **1601**, 1–32.

Gray, J. (1936). Studies in animal locomotion. VI. The propulsive powers of the dolphin. *J. Exp. Biol.* **13**, 192–199.

Hochachka, P. W. and Somero, G. N. (1971). Biochemical adaptation to the environment. *In* "Fish Physiology" (W. S. Hoar and D. J. Randall, eds.), Academic Press, New York and London, **6**, 99–156.

Hoerner, S. F. (1965). "Fluid Dynamic Drag", 2nd ed. Published by author, Brick Town, N.J.

Lighthill, M. J. (1970). Aquatic animal propulsion of high hydromechanical efficiency. *J. Fluid Mech.* **44**, 265–301.

Lighthill, M. J. (1971). Large amplitude elongated-body theory of fish locomotion. *Proc. R. Soc. Lond. B.* **179**, 125–138.

Tytler, P. (1969). Relationship between oxygen consumption and swimming speed in the haddock *Melanogrammus aeglefinus*. *Nature, Lond.* **221**, 274–275.

Wardle, C. S. (1975). Limit of fish swimming speed. *Nature, Lond.* **225**, 725–727.

Wardle, C. S. (1977). Effects of size on swimming speeds of fish. *In* "Scale Effects in Animal Locomotion" (T. J. Pedley, ed.), Academic Press, New York and London, Chapter 19, pp. 299–313.

Webb, P. W. (1975). Hydrodynamics and energetics of fish propulsion. *Fish. Res. Bd Can. Bulletin* **190**, 158pp.

Index

UNIV. COLL. N.W. BANGOR LIBRARY

£8-50p.
Mar Sci